MYSTERY DOWN UNDER

Murray Stuart

ISBN-13: 978-0-6459423-4-7
Cover design by: Currumbin Design
PUBLISHER: Good Beach Reads
www.goodbeachreads.com

Life is a symphony of riches
We dance to the music of the spheres
The universe of blazingly beautiful and brilliant souls

The photo on the front cover is of a sculpture "The Grand Impedimenta" by artist Tim Storrier. It is currently on display at the Tweed Regional Gallery in Murwillumbah, New South Wales, Australia.

Aboriginal rock art -Ubirr Art Site,
Kakadu National Park, Northern Territory.

Australia Creative Commons Wikimedia.

CHAPTER 1

Once upon a time. How can a story that covers more than 20,000 years start in any other way? Under the vast, starry skies above Australia, a girl named Yindi is gazing up at the constellations, her eyes trained on the Southern Cross. She is fascinated by the astronomical knowledge and practices of her people, and she is determined to learn all she can about their ancient wisdom from her relatives and community. She asks her grandmother, Kuralini, to tell her more stories about the stars, while the flickering light from the campfire does little to extinguish the brilliant luminosity above. The family turns their gaze skywards, watching the familiar constellations of stars that guide their seasonal journeys, hunting, and food gathering from the abundance offered by the natural resources of the terrain.

Yindi wants to know more about the two cockatoos flying through the sky chasing the Southern Cross. She asks if they are flying up to roost in the Ti-Tree stars hovering above the black hole in the sky. Her grandmother tells her that could well be the case. And does she know about the seven beautiful girls, one very shy, hiding behind the light of the others? They are looking at the handsome young men standing in a straight line. They tell them to keep their distance. Yindi wants to know where the Seven Sisters come from. Kuralini tells her they've always been there, in the stories told down the generations by long gone elders.

Long, long ago when the people would move along the coastlines, collecting food, staying at camp sites, slowly migrating, generation after generation, to new places, until they came to a new land with big emus, crocodiles and giant kangaroos.

The next morning, as they start walking along the beach heading for another of their camps, the timeless sea green ocean continues its ceaseless pounding of waves onto the beautiful white sandy shore, fringed by a lace of white surf.

CHAPTER 2

After a long day, Melanie pulls into her accommodation in Katherine, gathering her day's paperwork and an empty drink bottle, an apple core and some mandarin peel, places them in her laptop backpack, finds her shoes, and wanders into the studio apartment where she is staying overnight after a field trip from Darwin to a scheduled Northern Land Council meeting.

Melanie feels a deep sense of relief to arrive back, keen to unwind and find solace in the tranquillity of a private space. As she steps through the door, a wave of relief washes over her, as she leaves behind the stress and demands of the outside world, and the very real problems that people are facing in their daily lives.

She kicks off her shoes, feeling the cool, comforting sensation of bare feet against the wooden flooring. She takes a deep breath, a hint of lavender and a subtle lemon fragrance coming from the bathroom. She wanders around the studio, appreciating the aroma of fresh linen.

The soft ambient lighting bathes the careful minimalism of the space, casting a warm and inviting glow, illuminating the indigenous art that graces the walls. She sinks into a comfortable armchair, feeling its plush cushions support her in a comforting embrace. She leans back, allowing the tension in her body from the daily grind to gradually dissipate. After a few minutes, with a sigh she pulls herself upright, and moves to the kitchenette to make a cup of tea, returning to place it on a small side table. She takes a sip, enjoying the warmth, the gentle waft of blackberry adding to a sense of calm.

She indulges in a few moments of silence, revelling in the quiet of the surrounding bush. She closes her eyes, feeling a moment of peace. She thinks about reading a book, alternatively, perhaps a quick look at the news. She reaches for the remote and after a few minutes of the daily round of politics, accidents and foreign affairs, she turns the TV off again. The warm light from the floor lamps casts shadows creating an intimate atmosphere. She accesses the youtube app on her mobile, and selects an ambient music video, turning off the screen, and lowering the level to a few notches above audible. The sound of waves on the seashore gently fills the room, further melting away the stress of the day, an auditory backdrop suitable for a quiet evening.

Melanie glances out of the window, appreciating the changing colours as the last of another gorgeous Northern Territory sunsets graces the sky. She is captivated by the fading blue dissolving into the bright tropical hues of orange, pink, and purple, a reminder to enjoy the beauty of the natural environment, something that exists beyond the daily routine. Eventually, hunger asserting itself, she prepares a simple, nourishing meal from fresh ingredients assembled from a hasty supermarket shop on her way to the lodge. The aroma of a garlic, onion, and tomato omelette permeates the air, adding to the sense of domestic comfort and contentment.

Having enjoyed the food, Melanie settles down on

her couch to watch Evening Briefing on the national broadcaster's playback facility, iview, a segment that she had recorded the previous week, with Graham Saunders as interviewer. She'd only agreed because of her respect for Graham's interview skills - in her view he can be trusted to stick to the facts, and not to twist her words into another context.

In response to Graham's question about the effects on the surrounding country, she had replied. 'Uranium mining can have significant environmental impacts, including contamination of soil, water, and air. ERA's uranium mining has produced radioactive contamination. We are talking about a naturally occurring radioactive element. Mining and processing uranium releases radioactive dust into the environment. And during transport. Then there's polluted water and tailings, left over waste materials. '

'So, Melanie, what is the post mining legacy for Mirarr people? ' Graham asks her what she was hoping he'd ask.

'Uranium mining requires large amounts of water, which leads to the depletion of local water sources and often radioactive material enters the nearby waterways. Heavy metals, radioactive materials, and acid mine drainage can enter the creeks and affect aquatic ecosystems.'

'So you are saying that elevated levels of radium in the water could have widespread effects?' Graham throws her another Dorothy Dixer, a question she wants to be asked.

'Yes. Widespread water contamination is a distinct

possibility. And then there's the effects on soil. Uranium mining leaves soil with heavy metals and other pollutants, affecting the health of plants and animals and reducing fertility. Air pollution from dust and particulates is responsible for negative health impacts on people living nearby.'

'What is the community response to the rehabilitation of the mine?' Graham is always fair in presenting both sides of an argument. 'The Australian Government has secured the ongoing rehabilitation of the Ranger uranium mine by ERA, and the gradual return of the site to the Mirarr Traditional Owners, with continued government oversight of the clean-up. Surely this is a landmark rehabilitation project?'

'In my view, there has not been enough effort to minimise impacts and engage effectively with people living in the local communities. As you know, the Ranger uranium mine is located within the Kakadu National Park, operated by Energy Resources of Australia, which is majority-owned by Rio Tinto. '

The broadcast continues with a cultured voice over archive footage showing the building of the mine. 'The history of the Ranger mine dates back to the early 1970s, when the Australian government began exploring the area for uranium deposits. Over the years, Ranger has produced large amounts of uranium ore. However, the mine has faced a number of environmental and safety issues.

In 1981, a spill of radioactive tailings occurred, leading to the contamination of a nearby creek and the surrounding area. In response, the Australian government required the mine to implement environmental and safety measures. It has continued to face criticism and controversy over its environmental impact. The final shipment of uranium ore was sent in 2021, and the site is now in the process of being decommissioned and rehabilitated.'

The camera returns to the studio, where Graham is facing the camera.

'The indigenous people of the area surrounding the Ranger uranium mine have had mixed opinions about the decommissioning. The mine has left a significant legacy of pollution and contamination in the surrounding area. The rehabilitation plan includes measures to restore the area to its natural state.'

A native bird song trills out, insistent over the sound of waves on the shore. Using the remote to turn off the Smart TV, with one hand, Melanie picks up her mobile phone with her other, swiping to take the call.

'Melanie? It's Ben. We've had a report of a strange vehicle in Gunlom Falls carpark. As you know it's closed. I'm in a community training session for a couple of days. Jabiru thinks it's a community policing matter, because it is probably just a park visitor stopping to check out that it's really closed. They don't want to send anyone else out to investigate, because it's just one report. That's a four and a half hour round trip, and a lot of petrol, and it's probably a waste of time.'

'Yes, Gunlom Falls has been closed for months. There are signs up everywhere, and there's no access to the stairs. '

'I know you are in Katherine, and you'll be attending the Kakadu Board Meeting tomorrow. and I figured you'll be heading to Jabiru early tomorrow morning, and as you're driving back from Katherine. I take it the meeting's not till early afternoon, and then you'll be heading back to Darwin?'

'You have a Bush Telegraph at your cop shop?'

'Yes, Bob called in on his way back from Katherine. I was just checking in after the course finished. And he happened to mention that he was in a meeting with you today, and that you would be at the board meeting tomorrow. So I was wondering if you could take a short detour on your way tomorrow morning. Just swing by and have a quick look around, and see if there is still a car in the car park? I take it you'll be driving north on Highway One before turning off at Pine Creek to head for Jabiru? It's probably nothing, and it's only a short detour. After all, we can't be too careful with sacred sites.'

'OK Ben, I'll do that. I'll go there in the morning on my way. Actually, it's a good excuse to go see one of my favourite places'.

'Thanks Melanie. Be seeing you'.

'See you Ben.'

As the night deepens, Melanie gradually winds down, body and mind embracing the restful atmosphere. She climbs into bed. In the quiet solitude she reflects on the day's events, glad that, unlike in the past, landowners must now give informed consent before any action is taken to affect their country. She is gathering skills and learning valuable lessons on how to effectively negotiate with government officials in her role with the Northern Land Council. She jots down some preparatory notes on the board meeting including her concerns about the Kakadu National Park, the proposed uranium clean up negotiations. A sense of serenity envelops her. Her breathing slows, she relaxes, and her mind drifts off, letting go of the day, confident that tomorrow will take care of itself. Slumber overtakes her.

The following morning Melanie pulls into the car park at Gunlom Falls. No sign of any trouble, in fact no sign of anyone. She decides to take the walk up to the pool on the existing track, looking around to see what she can see, checking out for any further disturbance to the sacred site on the way.

On reaching the top, she stops. The sand coloured rocks frame a view across savannah grasses and scattered eucalypts to low rising hills. Looking across the pristine pools to where the water disappears over the edge into the distant landscape, Melanie spots a floating object at the far end, just where the water cascades down to the plain below. She banishes her first thought, that it is a croc, and waits for the beating of her heart to slow, because of course this is impossible. There have never been any crocodiles in the top pools. She distrusts the evidence of her senses. It must be a floating log.

Gunlom Falls has been closed since the trouble caused by the building of the new walkway. Parks Australia has since been prosecuted for these works undertaken near the sacred site next to the falls. No permission or consultation was sought, and the Jawoyn people, traditional owners of the country where the Falls are located, are incensed at the disturbance to a place of sacred men's business.

As she scrambles across the rocks, up and around to reach the farthest pool, she looks towards the object again. No. It is not a crocodile, and there is no-one around, and no tourists in the pool, because of the closure. Melanie makes her way across the water's edge, with a view to where the water cascades to the lower pool. A gasp escapes her, this is a person, face down, not moving, in fact very still indeed. Is it possible that someone is doing some kind of floating meditation? After all, the view is spectacular. The sandy red rocky hillside frames a distance view of trees rolling out to a horizon of hills gently rising out of a bed of sandstone. The stillness is almost peaceful.

Melanie moves closer to what is clearly a male, fully clothed, motionless, preternaturally still. In fact, face down and seemingly lifeless. She wades into the water, being careful not to go too close to the edge where the stream flows down to the plain below. She pushes and pulls the figure, and after several minutes of hard work, has pushed him up onto the rocks. She rolls him over so he is face up. Melanie is pretty sure he is dead, but she still checks. She cannot find a pulse.

Pulling out her phone, she dials Ben's number. 'Constable Ben Mooney, can I help you?'

Cuz, it's Melanie. I am at Gunlom. Up at the Falls. There is a man in the water. I pulled him out, but there is no pulse, and honestly, he is not healthy. He's dead.

'So who is he? Is he a blackfella?'

'No. He's white.'

'Do you know him?'

'He's not from around here', Melanie continues. 'And although his face is swollen, making it difficult to recognise, I have an awful feeling he is that bloke Parks Australia sent up from Canberra.'

'Just stay there, Melanie. I'll call the Sergeant at the station, and get back to you as soon as possible with an update.' A few short minutes later, Melanie's phone rings. 'They are sending a whole team out, a senior constable and first responders as well as forensics from Darwin. Just stay where you are. I'll be with you as soon as possible.'

'Thanks Ben. It's a bit creepy, and I have a board meeting to go to.'

'You don't have to stay up at the pool. Go back down to your car and lock yourself in. I am out in the field at Pine Creek, so I can get to you in half an hour. I reckon the Darwin lot will take at least four hours.'

'OK, see you soon'. Melanie ends the call, takes several deep breaths, looks around cautiously, and then starts the descent back to the car park, to wait for Ben.

CHAPTER 3

The Bright Green Coalition is elected to government. In fact, the Prime Minister of Australia, after a deal done with the many independents who were elected on preferences, is now, surprisingly, a young woman. The major political parties were decimated. This is a complete change for Australian politics. The old parties suffered a deep electoral defeat in favour of independents, including polarised fringe movements such as climate conspiracy theorists, legalise marijuana, men who want to be men, and monarchists. The BGC gets over the line due to a runaway youth vote, from a generation completely abandoning the politics of their parents.

When all is done and dusted, exit polls revealed a deep public backlash against increasing energy costs, the approval of new fracking mines, the declining standard of living, and a recession. Gen Z mostly voted for independent candidates because of a housing crisis causing a level of homelessness, the likes of which has not been seen in Australia since the Great Depression of the nineteen thirties, triggered by the 1929 crash on Wall Street. Australia was just as badly affected as the rest of the world, with large numbers of people walking from town to town carrying swags, looking for work. Many people lost their jobs, and struggled to afford basic necessities like food, housing, and clothing. As more people could not pay rent or mortgages, large camps of homeless people dotted Australian cities and towns. Young people feared a repeat of history.

The new PM's victory speech is emotional and emotive, emphasising the virtues of representation and honest government. In the election, a plethora of candidates ran as independents, with no party affiliations, each candidate presenting themselves based on their individual platforms, qualifications, and values. Voters were focused on addressing the real problems besetting communities up and down the country. Instead of voting along well worn party lines, for parties who failed to deliver promises for social benefits, innovation, economic growth, and positive new directions, the people voted for independent candidates across the board, while young Australia voted for the BGC.

People proved to be singularly weary of politicians doffing their caps to big business, while families suffer. Collectively, they deplored the constantly rising energy prices, rock bottom housing affordability and disturbing increases in domestic violence and homelessness, particularly affecting women. In a watershed moment, Australians decide that enough is enough. They've tried every variation of political party, left, right and centre over the past century. The green party is seen as irrelevant, entrenched in the existing tainted politics of a party system unable to provide a fresh perspective.

Following the national election, when no single party emerged to secure an absolute majority in the parliament, the BGC engaged in a coalition building exercise, crucial to forming a stable government, with newly elected MPs whose views are sympathetic to their own. This includes disaffected party members of all persuasions, now able to freely cross party lines in the wake of the election fiasco. In the end, many decide to throw in their lot with the Bright Green Coalition, seeing it as the only viable option to have any influence. The BGC proceeds to successfully negotiate an alliance to determine the policy agreements required to create a coalition government.

The new PM and her youthful inner circle decide on a consensus approach to decision-making, essentially accommodating the diverse views of the cross benches. Their promise to listen to the people who elected them, in the estimation of the BGC, is the single biggest factor that got them over the line. Youthful sincerity clearly impressed jaded voters, winning the day and reaping the benefit of the people putting the major parties last.

During the post election media commentary, the question on everyone's lips is 'why?'. Media views ranged from people's new found ability to distinguish political spin, to weariness of never having their questions answered directly by the major party politicians. In the wash up, the consensus is, that at heart, people were just tired of being played for mugs. When determining their party name, BGC was chosen as close enough to the beginning of the alphabet to be placed high on the ballot paper, with the aim of gaining the donkey vote that always results when the public are particularly disenchanted with politics. Young candidates stood as BGC, promising to vote independently. The movement grew rapidly as desperate renters, workers, unemployed, students, new graduates, pensioners, and low income earners lost all hope of getting a home loan approved by the banks, whom they are sure are short changing them. The promised prospect of green loans for sustainable housing is a ray of sunshine in the otherwise gloomy economic outlook.

Such is the fervour with which Gen Z planned to take over Canberra, that they find unlikely financial support from a significant cohort of older Australians. It is their grandparents, not their parents' generation, who pay for their candidacy and campaigns. Needless to say, they are largely conducted online with political savvy, deep social media knowledge, and very successfully. Many older people decide to bypass leaving money to their sons and daughters in their wills, in favour of providing financial support to the next generation, paying for election expenses. As one well known older media icon and national treasure declared, when asked as to why she was endorsing this movement 'Well, we were the generation that protested at the Vietnam War. We were young, we had ideals and the courage of our convictions. We know that climate change is the big issue, and these young people are the ones who have to live with the consequences. We are tired of our children putting money above all else, buying more houses for their rental income, instead of doing something to address the increasingly ferocious fires and the rise in widespread flooding events that are destroying our environment almost every year now. Something has to be done. And someone has to stop the fossil fuel industry from pumping more gases into the atmosphere. My grandchildren will have to do it, because the politicians clearly cannot.'

And so young people take up the challenge,

exercising their electoral muscle, to form the Bright Green Coalition of like minded people. They target schools, clubs, bars, universities, colleges and workplaces with clever, concerted social media posts. The movement started with the under twenty-fives, then was quickly adopted by every progressive sector of society. So many people want to see fair, socially inclusive policies implemented, not the will of the politicians. They want agreement and action on major social reform, not division funded by the rich and powerful.

To maintain the spirit of independence and prevent the concentration of power, the newly minted government MPs decide to implement a rotating leadership system. Each member can take turns holding key ministerial positions, including the crossbenchers, providing a variety of voices to lead and influence policymaking.

They immediately establish a proactive, participative social media platform called OzTalk, allowing people to engage with online forums, interacting not only with each other, but also directly with parliamentarians. Politicians have to make themselves available at least two days per week. This generates real proactive health, housing, education and economic policies, meeting the real needs of the population. And to further prove their democratic credentials, community communication channels for the elderly, disabled, and those without access to the internet, are set up everywhere across the country.

A local community volunteer outreach program provides practical help with online services, especially for older Australians, still challenged and baffled by the incomprehensible and badly programmed applications of government, banking, insurance, telecommunications, water, electricity and other essential services.

In the few short weeks after the election, these programs are up and running. Federal public servants are encouraged to get involved as part of their job description, with the veiled message that failure to do so would be a career dampener. Everyone is invited to lodge proposals for local renewable energy projects, as part of the new Decentralised Australian Grid. Affectionately known as the DAG, it is commissioned by the incoming government to upgrade the failing centralised monolithic state based grids designed a century previously. The Internet of Things is the great enabler, using local expertise, artificial intelligence and cloud computing to automate supply and demand. New storage technologies for renewable energy are sought to bypass the big battery problem. This is the perception that there is a scarcity of minerals, resulting in ruthless exploitation of people in developing countries. It is quickly discovered that there are plenty of alternatives.

The BGC has already engaged in deep thinking and research prior to the election. They use communication networks established during student days, gathering best practice information from all the available sources that were so studiously ignored by previous governments in favour of big business profit motives. Consultation makes it clear that it is not difficult to minimise demand for batteries, by using complementary technologies alongside solar and wind.

International reports reveal that rare earth minerals are not really widely used in solar energy and battery storage in any case. Despite their name, they aren't actually rare at all. Their criticality is mainly related to the current virtual monopoly of China for extraction and processing.

The business tax regime is adjusted to favour local renewable energy. Decentralisation makes it relatively easy to find solutions. Urban and rural communities near the coast install pods of tidal and wave generators in estuaries and inshore along the coastline, to meet demand when there is no wind or sun. Advanced geothermal storage is built for communities away from the coast. In the capital cities, local solar installations are backed up by hydrogen storage. Hydrogen and electric vehicles become the new normal for transport. Local garages are transformed into green transport stations. Local governments are encouraged to engage in community housing and energy proposals, with the availability of green loans from superannuation funds. They use tried and tested energy transitions from Scandinavian practical experience. No point in reinventing the wheel. A ban is placed on costly, inefficient international consultancies, replacing them with suitably experienced local businesses, creating new jobs.

All of these measures are quickly put in place after a bill declaring a National Climate Emergency is introduced into parliament, taxing super profits by wealthy corporations, fossil fuel companies, miners, banks and supermarkets. When the anticipated accusations of destroying business are levelled, the new government repeats its favourite media trope, 'time to give something back'. This proves very popular with the public. A follow up campaign effectively gets the message out with concerted media posts and blogs. 'These big companies have disrupted Australia's social equality and natural environment, taking far more than their fair share of profits.' The penny finally drops that the profits are mostly going offshore. Land rehabilitation, restoration of water sources, trees, native species habitat, and the resurrection of coastal marine ecosystems are the new government's top priority. The money raised is invested in technology innovation leading to good jobs, new apprenticeships and on-the-job training for young people, where they are needed, working with new clean tech, and out in the natural environment.

Climate change, renewable energy, and conservation of species threatened with extinction become the new targets for employment programs. National parks are extended to include semi urban areas surrounding cities. The government sits down with, and listens to aboriginal elders and traditional owners, as they've been invited to do so many times in the past. New funding for indigenous rangers reduces the impact of floods and fires, improving land management with traditional practices. Energy efficiency programs make the transition to renewable energy universal with low cost loans. All buildings, domestic and commercial, have to comply with new laws, ensuring that everyone is engaged in the energy transition. Renewable energy technology projects are rolled out by communities and local governments, sweetened by the super profits revenue.

Indigenous Australians regularly address parliament with views, solutions to problems and social change. Funding for Closing the Gap in communities, results in improvements to health, education and employment. For once, they are listened to.

Big companies threaten to leave the country, however with new international rules for supply chain carbon pricing, the government calculates that the cost of re-establishment would mean that most companies are more likely to stay put. This proves to be the case. The government no longer bows to corporate economic pressure because meaningless consumerism is replaced by recycling and emissions reduction.

Carbon pricing stops stagflation caused by international re-shoring and friend-shoring. On-shoring and competitively off-shoring the supply chain lowers costs. At last, Australia, in the company of an increasing number of countries, ensures people benefit from the profits from national infrastructure and resources. Local businesses and human capital are valued. This is indeed an exciting time in history, the start of the radical changes required to address the rapidly escalating risks of climate change.

New programs for social well being are established, combining a guaranteed basic income with green social housing. Jobs are offered under the new OzSkills umbrella. People of all ages and abilities are given opportunities to get involved in meaningful ways. Ageism, sexism and racism are addressed with affirmative action. The word 'welfare' and the unconscious language of social stigma are removed from all government websites.

Grace looks up from her laptop, from the text she's been reading aloud to her lunch companions out in the garden of the Art Gallery of New South Wales cafe. 'So that's it. That is basically the ChatGPT assisted scenario for our generation entering politics and changing the Australian government at the next election. Of course I worked on it to stop it sounding like it came out of an AI engine.'

'Good job!' says Malcolm with enthusiasm. 'Love it. And I really like the name Bright Green Coalition, it will be right up at the top of the ballot paper. Do you suppose that really makes a difference?'

'Probably does.' Eka gives her opinion. ' I know there is a big donkey vote at every election, people just tick the candidates near the top, because there are so many names. Well, great piece of fiction, Grace. But pigs will fly'.

Grace sighs deeply. 'Two chances, Buckley's and none' she adds to Eka's scepticism. 'So what are we going to do to stop the country from backsliding into giving more big tax breaks to the rich mining companies digging up coal and fracking more gas at Beetaloo Basin, sending it out to the world on carbon intensive transport from costly new port facilities in Darwin, funded by the taxpayer. OK, the government is becoming increasingly unpopular, and the opposition parties keep shooting themselves in the foot. Surveys show that support for the major political parties is still dropping. The poor old Greens must be sick of always being dismissed by the media and the major parties. They must be sooo sick of being treated as though they are uni students with silly idealistic proposals, even though they are mostly in their fifties now. '

'Well the two major political parties are ignorant about climate change. And the media are not much better. They seem to have no idea that Antarctic sea ice is now melting much faster than the Arctic. The danger of sea level rises grows every day. Their idea of climate solutions is to give money to the big miners and energy companies who make fake promises to do renewable energy by 2050. Tick in the box, job done. 'We will never do anything that will in any way jeopardise the economy. We are responsible economic managers.' ' Malcolm imitates the treasurer of Australia.

'Yes, and about as much use as the jobs mantra that the prime minister utters at every opportunity. When anyone with half a brain knows that there are only a handful of jobs in fracking, and almost all the money from mining coal and gas goes off shore to American, German and UK companies. I've even heard that Rusal is still in a secret partnership with Putin and his mates. Time for me to leave. I better get back to work.' Grace looks at the time on her phone, and makes a move to rise from the table. 'Oh by the way, there's a new study out that says that the Gulf Stream will shut off anytime between next year and the end of the century, most likely around 2050. And of course every ton of CO2 from burning fossil fuel brings the date forward. ' adds Eka.

'Depressing.'

Grace and her companions exchange farewells as they wander back to work, crossing the Domain to the City of Sydney. Back in the office at the Australian Museum, Grace decides that she prefers the ChatGPT scenario as more comprehensive, interesting, and transformative to any political spin. The major parties seem to be the lackeys of coal and gas miners, providing the same old greenwashing she's been hearing since she left school, the latest being a massive propaganda exercise on Carbon Capture and Storage, essentially an end-of-pipe waste management response that does not reduce emissions. This amazing feat of self delusion is echoing around the world.

'What kind of future is there for people of my generation?' she wonders.

CHAPTER 4

There is a sense of calm on the Swiss side of the Rhone Valley, Majestic alpine country towers above the picturesque landscape, dotted with chalets and bouquets of greenery from the trees and shrubs that cover the lower slopes . The air feels crisp and invigorating, as if it carries the essence of untouched wilderness from the high country. Into this facade of tranquillity, a thunderclap is suddenly released from the dark grey piles of cloud hovering over the distant Mont Blanc massif, and a bolt of lightning accompanies the start of a powerful mountain storm, about to unleash its fury, not only on the terrain of the French Alps, but also moving rapidly across the Rhone Valley, carved out by an ancient glacier, towards Switzerland. Local people, looking across from the Ormonts Valley to France, take one look at the cloud front, and leap into action, rounding up loose objects, garden implements and children's toys, closing open windows and doors, and moving the cars, for those who are lucky enough to have undercover parking. It is clear from the height of the mass of dark clouds spilling from above three thousand metres, that this is going to be a storm on a grand scale.

As the day progresses, the temperature drops gradually, and the atmosphere becomes charged with a palpable tension. Clouds, initially fluffy and white are gathering on the horizon, painting a beautiful contrast against the deep blue sky. As the morning wears on, these innocent-looking clouds have morphed into the dark, towering behemoths looming over the valleys and ridges, moving rapidly to the Alpes Vaudoises.

By early afternoon, more distant rumbles of thunder echo through the mountain range, a warning sign of an impending storm. The wind picks up speed, carrying with it a biting coldness that chills to the bone. Gusts sweep down the slopes, rustling the trees and whisking away loose debris. The once serene scene transforms into a foreboding and ominous display of nature's might.

As the storm approaches, the clouds merge into a thick, grey blanket that obscures the towering peaks. Lightning flickers within the dense mass, illuminating the darkness with momentary bursts of blinding light. The roar of thunder grows louder and more frequent, a symphonic crescendo reverberating through the valleys.

The initial spatter of rain doubles its periodicity exponentially, at first sporadically and then in a torrential downpour. The rain's intensity varies over the hours, at times a gentle mist that envelops the landscape, and then a deluge, threatening to flood the narrow streams and rivers that wind their way through the deep gorge formed during past ice ages. When the glaciers last receded, rivers carved notches into the floors of what were once glacial valleys, deeply scouring the landscape.

At higher altitudes, the rain turns to snow, coating the mountaintops in a pristine white topping of what looks like sprinkles of icing sugar. The snowflakes dance and swirl in the powerful gusts, creating an ethereal spectacle. A beautiful perception that comes at a cost, as the storm poses risks to people and wildlife alike, making navigation on the roads treacherous, and challenging the survival of any living being caught without shelter.

Visibility diminished, the outlook beyond a few metres became mysteriously hazy with curtains of rain. With the storm's fury at its peak, for a time nature seems to reclaim the mountains, reminding all who venture into its domain of real elemental power.

But this is no ordinary storm, people are calling it a tornado, something that has not happened in living memory. The severe downburst is a prelude to a supercell thunderstorm, the type of extreme weather event that occurs because of an energy differential between cloud layers. The air cools, compresses, and further accelerates the downdraft caused by the falling drops, accelerating the strength and ferocity of the storm below. Although never before in the meteorological record, the ferocity of the storm is in accord with climate change predictions for a new intensity of weather events.

The storm passes quickly enough, but the strong winds have caused significant damage. The clouds begin to disperse, and the rain and snow recede, revealing once again the awe-inspiring beauty of the Swiss Alps. The air is now crisp and clear, and the sun begins to break through the clouds, casting a warm golden glow on the wet landscape, revealing uprooted trees, overturned cars, buildings damaged and trains derailed.

Switzerland is once again reminded of the raw, untamed power of nature, an experience that etches itself into the memory of all those who witness its frightening grandeur. In accord with weather predictions for increasingly hot summers, storms and droughts, Swiss and other European glaciers are well on the way to disappearing entirely. Local authorities are wondering whether they can create reservoirs to capture the water, once permanently locked up in the ice, before it flows away forever through the river systems into the oceans.

Despite the landslides and rockfalls, extreme summer temperatures and increases in the intensity of the mountain storms, Swiss people are doing what they've done for thousands of years of living with the extremes of climate that the mountains have always delivered. They are looking adversity squarely in the face, with stores of courage, and a steely determination to find strategic solutions to the new turn of events.

Fabian is influenced by his country's new climate predicament when he decides to pursue a career with the United Nations in Geneva, after finishing his studies. He decides he wants to make a difference, to protect the natural environment. He ends up with a coveted job in UNESCO, New York.

CHAPTER 5

According to environmental research, Australia is left over from the great southern land that began to break apart over five hundred million years ago, leaving in its wake, the original rainforest, stuck to the crust in the southern hemisphere, while most of Earth's land mass drifted ever northward. Down under, where the Southern Cross points to the south and the visible stars and moon look reversed to those who come from the north, Gondwanaland, the ancient supercontinent, slowly developed its flora and fauna, originally almost completely covered by ferns, conifers and angiosperms of ancient order.

The transformation of the landscape into the vast tracts of arid land beyond the coastal areas occurred relatively recently. The change happened some time after humans arrived sixty five thousand odd years ago, island hopping and walking land bridges from Indonesia and New Guinea. Once the balance was disturbed by people hunting the megafauna to extinction, vegetation was changed forever, with massive fire events started by lightning destroying vast tracts of forest. Water recycling over the interior of the land stopped almost altogether, and the rainforest retreated to tiny remnant areas as the rainfall dried up, and central Australia became hot, arid and sunny. This was millennia before the end of the last ice age inundated the coastline for hundreds of kilometres. A long time before the sea levels inexorably rose with the great melting event that occurred everywhere on Planet

Earth. Twelve thousand years ago melting glaciers drowned large stretches of the coastline. The Australian continental shelf submerged. The Great Barrier Reef became covered with coral.

The Northern Territory, a vast and diverse region, encompasses large groundwater systems and water tables, deep below the surface. Aquifers are underground layers of porous rock and sediment, the primary source of groundwater in the region. Many indigenous communities rely on groundwater. They are dependent on good water management to ensure access to clean and sustainable water sources. The water table is extensive, covering half of Northern Australia.

Mining disturbances caused by fleets of giant trucks and excavators, often comprise the wholesale clearing of trees and animals, further endangering threatened species, disrupting and endangering local water supplies. It is also responsible for destruction of Australian sites of cultural significance, centres for the balance of health and wellbeing of the natural environment. Juukan Gorge is a ravine in the Hamersley Range in the Pilbara region of Western Australia. International mining giant, Rio Tinto Group, in the pursuit of iron ore, first received ministerial consent to damage the site in 2013, culminating in the deliberate blowing up of caves populated for 50,000 years, since the middle of the last Ice Age. They were continuously inhabited by the traditional owners, Puutu Kunti Kurrama and Pinikura people.

The Juukan gorge is set amongst high and rugged ranges, an essential tributary of local rivers, and a corridor for wildlife to traverse, as well as a place of tremendous antiquity, with rockshelters carved out of the ironstone by countless millennia of natural weathering. Each of the rockshelters held a museum of information, holding thousands of artefacts, grinding stones, rock seats, blade quarry and flakes of stone from the people who used the caves to manufacture tools for hunting and fishing. The caves included human remains of ancestors of the local people, all destroyed in a moment of madness. No one could persuade Rio Tinto to halt this incredible act of cultural vandalism. They valued the money from the ore too highly. No wonder the archaeologists are still shaking their heads in disbelief.

Melanie is returning Fabian's call. 'I hope this time slot is OK. I am on my way to work on Monday morning. I am hoping this is a reasonable time to call.'

'Just fine, Melanie, it's early Sunday evening'

'Oops sorry, I forgot it is your weekend. I want to tell you what I have just found out about a new danger to ancient archaeology, and our sacred sites at Middle Arm.'

'I am always glad to talk to you. So what's happening at Middle Arm? Is that inside the perimeter of Kakadu National Park?'

'No, it's in Darwin. But it will have a huge effect on the Park's ecosystems. They want to build a huge gas facility. They're looking for environmental planning approval, right on top of sacred sites on Larrakia country. Ancient Larrakia rock art and shell middens that survived colonisation at Middle Arm could easily be destroyed by the gas hub. '

'Not after the destruction of Juukan Gorge, surely. '

'Yes. There's been no consultation with Traditional Owners by either the Northern Territory or Federal Governments. They are calling this development 'green', all about 'clean energy industries' and 'environmentally sustainable manufacturing'. '

Fabian can hear the quotation marks in Melanie's voice. 'Go on, tell me the worst'.

'Development is all about subsidising fossil fuels. The government has announced a handout to the tune of $1.5 billion for the Middle Arm 'Sustainable Development Precinct' in the port of Darwin. Taxpayer subsidies for fossil fuel companies are already at $10bn a year, if you include fuel subsidies. And despite all the greenwashing and green spin over the development of a second port for Darwin, they just give more money to the fossil fuel industry. They are pushing it as a world beating gas, hydrogen and minerals processing and export precinct. The plans include a petrochemicals plant for plastic, pesticide and fertiliser production. Green hydrogen is probably just green spin. '

'That's pretty awful, Melanie.' There is real concern in Fabian's voice. 'Surely that comes with a huge carbon footprint.'

'No doubt. The truth is that the second port development is there to increase trade activities, with no thought to the consequences. The Northern Territory wants to be a gateway to Asia. There are capacity and infrastructure issues at the existing port. Nothing to do with sustainable development. Middle Arm is projected to have deep-water berths, container terminals, and bulk handling facilities to attract major shipping lines. They are selling it as access to export markets, especially for minerals, agriculture, and, of course, liquefied natural gas. The environmental damage would be huge. The politicians are barely paying lip service to climate change.'

'Don't you have a new government? Didn't they get elected promising no new private investment in fossil fuel projects? And definitely no government money?'

'Yes. Complete hypocrisy. Climate Minister James Cowan said the government would not stop new private investment in fossil fuel projects but promised there would be no government money for new fossil fuel developments. But that turns out to be political spin. The reality is completely different.'

'And?'

The reality is the Middle Arm development will be used to finance the development of fracking in the Beetaloo Basin. Gas will be used to power manufacturing petrochemicals, and they want new LNG plants to turn gas into liquids for export. There is a commitment to huge private and public investment in CCS, carbon capture and storage, which they claim is clean and green.'

'But undersea CCS has been discredited as not being commercially viable. And it is environmentally damaging, not to mention unstable.' Fabian knows a lot about clean technology.

'Exactly. This project is all about gas. It is not green at all. Unless they've solved all the problems in the last couple of months?' Melanie poses the question in a rhetorical tone.

'No they have not. And yes, the carbon footprint is huge, requiring massive amounts of fossil fuel energy, and the technology is not currently practical at scale. There is huge doubt that it will ever be economically viable.'

Melanie continues 'Well the new government keeps talking up green hydrogen. There are huge questions over whether the green hydrogen promises will be met. Will it be green hydrogen or brown? Hydrogen is not suitable as a grid renewable energy generator.'

There is concern in Fabian's voice as he continues. 'Hmm. Melanie, this is not good news. CCS projects are impossible to finance without subsidies, and they require huge infrastructure, including pipelines for transportation and suitable storage sites. Establishing this on a large scale would be complex, very time-consuming, and of course, emit massive amounts of greenhouse gases. Don't tell me the new politicians support this madness, at a time when the global climate crisis is moving to catastrophe status, with both the Antarctic and Greenland ice sheets melting ever faster. '

'I know. It's really bad. If it goes ahead, it'll be decades before the carbon footprint is offset, even with the dodgy carbon credits and tricky carbon accounting. The potential for leakage of stored CO2 is massive, not to mention the impact on fragile country and the seabed. Flawed environment impact assessments are just the beginning of the problems.The environment minister, Anna Grasski, looks like she has been told to approve the project.'

'By whom?'

'By the prime minister. The party's largest donors are the fossil fuel industry.'

'So what is the government telling the public about this?' Fabian poses the obvious question.

'Oh, they are hoping that the tired old mantra 'jobs, jobs, jobs' will stop the public from questioning their green credentials. Every time I hear that phrase, I can't help thinking about an interview I saw with Greta Thunberg. When she was asked about the argument that gas development creates lots of jobs, she could not stop giggling. She told the interviewer that 'jobs, jobs, jobs' is a standing joke amongst her friends.'

'Yes. Well there might be a few hundred jobs, nowhere near as many as would be lost from ecotourism and agribusiness. The level of industrial pollution on the pristine environment is unthinkable. Not to mention the collateral damage from the bushfires and the floods caused by global heating. Middle Arm is not in the Kakadu National Park, so it is not officially part of my remit at UNESCO, however it is close enough for me to justify an initial exploration of what this means to Kakadu. Melanie, I'm coming to

Darwin.'

'Really Fabian, that's great! When?'.

'As soon as I can manage it. I've been following the debate over the Voice, and the shocking incarceration rates of indigenous kids. I am wondering if this is a smokescreen, to cover up for their lack of action on protection of sacred sites from fracking and mining? I'll call you as soon as I've got the dates for my flights. Can you help me find some accommodation? '

'I can do better than that, Fabian, you can stay at mine. I have a spare room.'

'Thanks Melanie, that'll make it easier to get the budget approved. If it's just for flights, I can get the process moving quickly.'

Melanie is smiling to herself as she gets into her car to drive to Mindil University. She has been asked to speak on the first day of the inaugural paleo astronomy course being sponsored by Melbourne University. Fabian coming to Darwin? Great!

CHAPTER 6

In Norway, in the far north beyond Trondheim and Tromso, is the proverbial endless fractal Norwegian coastline, where inlets and fjords frill the edge of majestic mountains with cascades of rocky terrain tumbling down to the water's edge. The mountains of Lofoten offer some of the most incredible scenery on the planet. Picturesque islands of fishing communities and farms border the sea's edge, surrounded by towering peaks.

This is the northernmost part of Norway, known for its unique and diverse landscape. To the east is the county of Finnmark, situated above the Arctic Circle, Norway's largest in terms of land area with, however, a low population density. Sami people have lived in Finnmark for at least 10,000 years. The name derives from Old Norse. 'Finn' was the Norse name for Sami people. This was the name given to any places where Sami people lived.

Finnmark has stunning fjords that cut deep into the rugged coastline, surrounded by steep cliffs and mountains. Inland from the fjords, there are vast plateaus and expansive tundra landscapes, covered with low vegetation, and dotted with lakes, rivers, and marshes. Though largely snow covered in winter, during the summer months, the land is transformed with vibrant green vegetation. The mountains on the border with Finland offer recreational opportunities for hiking, skiing, and other outdoor activities, popular with Norwegians and visitors. With coastal foreshores, and several large islands off the far northern coast, ancient fishing villages abound, inhabited since the Bronze Age, with a deep tradition of art, crafts, song and the Sami tradition of Yoik. Yoik is the native Sami musical expression that provides a living connection to people and places in song, perhaps in some ways similar to the songlines of indigenous Australians.

The beautiful seascapes provide wide vistas northwards across the Norwegian Sea to Svalbard, the island chain halfway to the North Pole. Northern Norway is an excellent place to witness the beautiful dancing waves of light that have captivated people for millennia during the winter months, provided the weather is clear and conditions are right. The Northern Lights, the Aurora Borealis, are the result of disturbances in Earth's magnetosphere caused by solar winds, a stream of charged particles released from the upper atmosphere of the Sun.

The region is abundant in wildlife, including reindeer, Arctic foxes, and various bird species. Along the coast, there are sea eagles, seals, dolphins, and whales in the surrounding waters. In Spring, the reindeer travel from their overwintering to traditional coastal pastures, guided by Sami herders. They either swim, or are transported in trucks across fjords to reach the good summer grazing reserved for them. They are accustomed to human beings, and remain largely untroubled unless approached, constantly calmly moving across the land in search of tender shoots and succulent grasses, oblivious to the spectacular ocean scenery around them.

Northern Norway is surprisingly warm for its

geographical location. This is thanks to the Gulf Stream, part of the Atlantic Meridional Overturning Circulation. The AMOC is a crucial component of the global ocean circulation system, and it plays a significant role in regulating climate patterns in the Arctic Circle, including Siberia, Northern Europe, and North America. Dubbed the Atlantic Conveyor Belt, it is a large-scale circulation system that involves the movement of warm, salty surface water from the equator towards the North Atlantic. As the warm water flows northward, it releases heat into the atmosphere, meaning the temperatures in the regions it passes through are kept relatively warm.

For a comfortable lifestyle in Northern Norway, it is essential to keep this system running. In recent years the AMOC has shown significant signs of slowing down. This is the system of ocean currents that bring warm surface water from Africa northwards, and cold water in the depths from Iceland and Greenland south, creating a conveyor belt of warmth in cold weather countries, while facilitating African rainfall. When the warm water reaches the north, it cools down, becomes denser, and eventually sinks to deeper ocean layers. This deep water then flows southward along the ocean floor, completing the circulation loop. The process is driven by differences in water temperature and salinity, driving the current onwards. Nobody really knows what will happen in the short term with a slowdown, let alone a shutdown of this essential climate system. Presumably, as happened 13,000 years ago, northern Europe would ice over again.

This system also helps to moderate temperatures

all over the world, leading to milder winters in the north, and cooler summers in the south. It is a system of currents that is deeply interconnected with all the global circulation systems in the atmosphere as well as the oceans. It influences sea levels, the distribution of marine species, and weather such as the amount of rain and the severity of storms across the Atlantic. Changes will certainly have cascading effects on weather systems.

The Greenland ice sheet is melting at an unprecedented rate. In fact an astonishing 10,000 tons per second of ice, on average, melts into the ocean. And in recent years, there has been little replacement by the winter snows. The increased freshwater input from melting ice reduces the salinity of surface waters in the North Atlantic, making the seawater less dense and less likely to sink. The influx of massive amounts of freshwater into the North Atlantic is what has caused past shutdowns, leading to ice age onset, causing global climate feedback loops. These feedbacks influence other climate tipping points, such as weather systems and sea-level rise along coastlines, rainforest dieback in the Amazon, more forest loss, further collapse of the West Antarctic ice sheet, and of course ocean acidification.

Jonas, living on the island of Skjervoy, in northern Norway, reads the Guardian newspaper online, a habit he picked up at University. 'Climate scientists have detected warning signs of the collapse of the Gulf Stream, one of the planet's main potential tipping points. The Atlantic Meridional Overturning Circulation currents are already at their slowest point in at least 1600 years. The new analysis shows they may be nearing a shutdown as early as 2025.' He immediately sends a group email to his closest friends including the link to the article, with a subject line in capital letters LES DETTE!!**??**!! - read this.

He loves marine biology. His focus is marine ecosystems, and the interactions between them. Norway's extensive coastline and thriving industries based on the sea have given him plenty of opportunities to get involved in marine research and conservation during student days at the University of Bergen. With the challenges of warming waters around Norway's coastline, his coursework entails not only research into marine ecosystems,but also the impact of human activities on the fjords, and the rivers and streams that feed them.

Jonas has been able to engage in a number of projects monitoring the health and abundance of marine species and ecosystems along the coastline and offshore in the North Sea. Like his fellow students, he loves the fjords, the deep, narrow inlets carved by glaciers in the Earth's last ice age. He takes great pleasure in being out on a boat, looking up at the sheer cliffs that surround the fjords rising dramatically from the crystal clear waters, covered in dense spruce and pine trees. He participated in a monitoring program to assess the status of fish populations, marine mammals and seabirds, where the steep mountains along the coastline of northern Norway, meet the tranquil beauty of the sea. He is really happy to have a career unfolding where he can spend a lot of time outdoors, experiencing nature, while engaging in conservation efforts.

He has already gained experience in fisheries

management, sustainable oceans and looks forward to a career in the 'blue economy' that is a Norwegian government priority. He just finished engaging in a project assessing the impacts of offshore oil and gas exploration on shipping and aquaculture on Skjervoy, his island home. He loved being involved in ensuring and improving the sustainability of salmon farming in fjords and sheltered coastal areas. Though he doesn't really like fish farms, he knows that the oceans are being fished out.

Skjervoy is a beautiful island with an abundance of nature walks, spectacular views of towering mountains, dramatic peaks and awesome escarpments leading down to the waters of the fjords below. He really enjoys spending time on the island, particularly because of the chance to show the place off to his colleagues from the south, driving them around, fishing off jetties and rocks, and taking them across in ferries to neighbouring islands to see the migrating reindeer. He loves their soft woolly appearance and gentle countenances, and their lack of concern around people and villages, no doubt because of their long relationship with the Sami reindeer herders, over thousands of years.

He tells his colleagues the story of German occupation in World War Two, when locals were made to move out of the area, and their houses were burnt down. It was the treatment of the local Sami people, forced to move from their villages, with nowhere to go, and the Russian prisoners-of-war that were truly terrible. The Russians were forced to live in huts made of turf and birch branches that couldn't keep out the rain and snow, the prisoners were fed starvation diets and forced to work in terrible conditions, building roads for the Germans. Most died. A few local Norwegian farmers were allowed to continue to stay on their land, to feed the Germans. One story he knows about is a Russian who was sent daily to pick up eggs, secretly given potatoes by a local farmer to supplement the meagre rations.

Not faring much better were the local Sami people, who were treated as less than human. The Germans burnt their houses and settlements, and they also burned the forests. Without the forest areas, having much of the land scorched and burned, reindeer herding was severely affected. So, not only did the Sami lose their homes but also their way of life and the means of making a living to keep their families fed.

'Truly a terrible time', thinks Jonas, 'and so recent, in my grandfather's lifetime'. It makes him more determined to make a difference in his life, and not to let injustice go without at least trying to say or do something. He has a kind heart and deep sense of responsibility that does him credit.

CHAPTER 7

'Australian indigenous astronomy is a fascinating and important field of knowledge and practice of the first Australians. People have a rich history of observing the stars, planets, and other celestial objects. Complex astronomical systems developed over thousands of years.

Learn about the stars of the first Australians, to gain a deeper understanding and appreciation of traditional cultures and the natural world'.

Mara is interested in finding links to the stargazing of the ancients, something that has become a hot topic in European circles. She reads on. 'We offer a short course, Introduction to Paleo Astronomy, at Mindil University in the Northern Territory. It covers a range of topics related to the astronomy of Indigenous Australians throughout history. The course is interdisciplinary in nature, drawing on additional fields such as anthropology, archaeology, and indigenous art.

Students are offered a compelling range of topics, both on campus, based in Darwin, and out in the field in the Northern Territory traditional lands of the Yolngnu, Warlpiri, Mirarr and Anangu people. '

Mara is taking unpaid leave from her role at the Edinburgh University, to do some preliminary research about palaeolithic Europe. She has been reading about the possible star maps in cave paintings alongside the famous Lascaux art made by European hunter gatherers over 17,000 years ago. She finds the offer of

Australian palaeolithic astronomy intriguing. Thinking she can compare ancient European and Australian traditions, Mara reads on.

'Introduction to Australian Indigenous Astronomy: An overview of the different systems and practices of the First Australians throughout history, including the Dreamtime stories and Songlines that explain the movements of the stars and planets. Field trip to Kakadu National Park, watching the stars at night with local knowledge keepers, a chance to view ancient rock art, and study astronomical alignments.

Explore the cultural significance of the stars, including its role in ceremony, ritual, and everyday life. Delve into the observation and study of the stars and planets, the use of the sky for navigation, timekeeping, and seasonal cycles. The course features guest lectures from indigenous knowledge keepers and researchers, providing students with a unique and valuable perspective on this rich and complex field.'

'That's for me!' A colleague sitting adjacent looks up. Mara becomes aware she has spoken out loud at her desk in Edinburgh. 'You've found a new job?' joshed Dan, a fellow lecturer at the university, specialising in astrophysics.

'Yes, I've been offered a long term position with NASA.' Mara pauses for effect. 'Not. As you well know, because I've been here, goodness, for 10 years, I am taking a short sabbatical to write a paper on the astronomical alignments of neolithic Aberdeenshire stone circles. And, now I've seen this offer in Discover magazine, which fits perfectly with my research. Listen to this, Dan.

Two weeks Paleoastronomy Course at Mindil University, Darwin. Combination of classroom lectures with indigenous guest lecturers plus an unforgettable guided tour to experience the Kakadu World Heritage listed Cultural Landscape sites led by Traditional Owners. '

'Oh, I want to go too. take me with you.' There is no trace of irony in Dan's voice.

Mara continues. 'Yes. Sounds fantastic. The course is scheduled for next month. I can't wait. I've decided to spend the first part of my three months leave in Australia!'

With a sense of freedom to do what she likes, when she likes, wherever she finds herself, she fills in the application, pays for the course online, and begins to make plans for July, a month ahead, when the two weeks course begins.

'After the course, I'll do what I've always wanted to do, and spend a couple of weeks exploring what I can in a short time, ' Mara promises herself. 'Australia! Exciting.'

Search for flights reveals that she could travel British Airways from London to Sydney and Qantas from Sydney to Darwin. So the next decision is to fly from Edinburgh or take the train. She opts for the train in consideration of her carbon footprint. With the itinerary finalised, she walks into the kitchenette adjacent to the work space. 'What have I done?' she wonders, but before the doubts kick in, she pours a strong coffee from the cafetiere, and congratulates herself for her decision to do something different, and break out of her daily routine in a spectacular way.

She has been reading about the similarities between Australian indigenous and western astronomy, astonished to learn that the first Australians also called the constellation Pleiades, the Seven Sisters. The Pleiades is visible from virtually every part of the globe. It's seen from as far north as the North Pole and farther south than the southernmost tip of South America. It is a small misty cluster. There are astronomy stories and global myths about 'seven sisters' stars that may reach back 100,000 years. She cannot wait for the course to begin.

Mara's flight to Australia, lasting over 24 hours, proves a veritable long haul. It is a wearying experience indeed. She is glad when her taxi ride deposits her in front of a gracious older house in the beachside suburb of Fanny Bay. She rests up for a couple of days, doing as little as possible, as her arrival is followed by several nights of jetlag, waking up impossibly early, at three and four successively.

The house is one of the remaining old Queenslanders, run as an Airbnb. The style was often built in Darwin, inspired by traditional Queensland architecture, where houses are designed to catch the breezes and beat the summer heat. These houses typically feature a raised design, with living areas on the upper level and space underneath for parking, storage, or outdoor activities. With wide verandas that provide shade and airflow, high ceilings, and timber construction to promote natural cooling, they are practical and sensible for a hot weather climate. On the third day, Sunday, she wakes at a respectable hour, and promptly returns to bed to read, check her mails and generally laze about.

After a decent sleep and a good lie in, Mara sets out to explore the neighbourhood in the warm afternoon. She wanders aimlessly through suburban streets, vaguely heading for the beach. Crossing a road, she sees a tropical palm fringed site filled with colourful market stalls, people milling and wending their way leisurely along the rows of offerings, primarily food, ranging from local and international cuisines, as well as bright, creative clothing, natural therapy products and an array of arts, crafts and services. She stops to watch, and is entertained by a number of theatrical buskers showcasing their talent with a wide variety of musical styles.

Wandering past an array of tempting aromas, she finally settles on a crab laksa, and heads to the nearby beach, to picnic on the sand, and enjoy the Darwin winter sunset. The city is located in the tropical zone, and doesn't experience cold temperatures. Instead, winters are milder and drier, compared to the wet, the most notable weather season.

As the day wears on, the stunning display of sunset colours against the backdrop of the coastal views visible from Mindil Beach is just the type of experience Mara was hoping for, when she decided to make the trip. As the sun begins its descent towards the horizon, the sky colour gradually transitions into shades of orange, gold, and purple, with hints of pink and crimson. The refractive clouds create a delicate contrast with the fading blue of the sea.

'Note to self. Always make snap decisions to go to

stunning places.' Mara had no idea of what to expect before she headed down under. In Darwin, the sun sets over a wide horizon, casting its final golden rays across the sparkling waters of the Timor Sea. The reflections on the ocean creates a spectacle, and as the sun disappears, the water surface takes on the appearance of liquid gold.

The winter sunset in Darwin is a time when locals and visitors alike are outdoors, gathering at spots like Mindil Beach, to enjoy the markets, relax and appreciate the beauty of the natural world. In the last of the fast fading light, the silhouettes of palm trees, mangroves, and other tropical vegetation are shadowed against the slowly darkening sky. The foliage takes on deep green and black hues, accentuating the contrast with the brilliantly coloured backdrop, as the tropical night descends. The air is slightly cooler than during the day. The tropical evening is interspersed with the distant calls of birds returning to their nests for the night, and the gentle lapping of waves against the shore, as the tide comes slowly in.

Mara is glad that she made the decision to go halfway around the world, and retires for the night, having written her impressions in her diary. She heads for sleep, happy to have travelled to such a new, different and exciting place, happy with life in general.

Emerging from her accommodation the next morning, fully refreshed from a relaxing weekend, and the visit to the Mindil market, she sets out on the short trip to the campus of Mindil University, guided by google maps.

Finding her way to the appointed seminar

room, she finds herself seated next to Grace, a woman archaeologist from Sydney. They strike up a conversation and quickly discover they share a passion for indigenous culture and history. Grace has been studying the rock art and ancient artefacts created by early first Australians, and has uncovered evidence of astronomical knowledge and practices. She is keen to learn more about their perceptions, their reality and their experience under southern skies.

'So what actually motivated you to sign up for this course, Grace?' Mara is genuinely interested and curious about the clearly interesting young woman sitting beside her.

'Oh well, I find it amazing that people here have had a deep relationship with the stars for tens of thousands of years. I have always loved looking up at the Southern Cross. I was really surprised when I learnt it cannot be seen in Europe. I know nothing about astronomy, so I thought this would be a good place to start learning.'

'So what kind of work do you do?' Mara is curious now.

'I'm an archaeologist. At the Australian Museum in Sydney. I want to know what the first Australian people were doing and thinking for the last twenty thousand years since the sea levels rose after the last ice age. I find it absolutely amazing that life before the sea flooded inland along the Australian coast is still told as stories, and sung as songs right up to today. Aboriginal people have preserved their oral histories in a way nobody else on the planet has done.'

Mara decides to try to explain the difference between

the stars of the northern and southern hemisphere to Grace. 'You do have a unique view down under. In Scotland, the night sky looks different because of our latitude, and our lines of sight as observers. Different sets of constellations are visible. Like Orion - you have to look toward the south to see Orion in the northern hemisphere, while people in the south have to look to the north to see Orion's belt, and the stars appear to be flipped from left to right. The Southern Cross, visible in Australia, is invisible in the north. Well, it has been invisible since the time of the ancient Greeks, who could just see it above the horizon. That's due to the curvature of the Earth. Then the position of the Earth's axis, and the orientation of its orbit relative to the stars, has changed over time. This is because of something called the precession of the equinoxes, a slow shift in our position along the ecliptic, the plane of Earth's journey around the sun. Think of a spinning top slowly wobbling around its axis every 26,000 years, due to galaxy gravity. The same reason why the pole stars, the ones that appear stationary over the North and South Poles, have changed over the last few thousand years.'

'Wobbling like a top every 26,000 years. I think I have a very vague idea of what you are saying. I clearly have a lot to learn. I'd like to get in touch with people who have used the alignments of stars for practical purposes since forever. For me, archaeology is not just about old sites, pottery and rocks, it is about the people who lived in those places. And what they were thinking, and how they were living, and how much they are just like us. And what we can learn from their daily lives. That is why I am on this course. To understand how we all used astronomy in our daily lives once, because we were all hunter gatherers.' Grace realises that she's just said what she's been thinking to a complete stranger. ' You said you were interested in indigenous astronomy, does that mean indigenous Australian?'

'Actually, no. This is my first exposure.' was the reply from Mara. ' I've had a chance to do some research on the astronomy practised by the Sami people of Northern Norway. And that is fascinating. Sami culture traditionally follows a seasonal calendar closely linked to celestial events. Events such as the positions of specific stars or the appearance of the Northern Lights, associated with the changing of seasons. Or the timing of certain activities, like reindeer herding or migration to another location. I want to find out whether there are any similarities.'

Mara and Grace are immediately comfortable in each other's company, and are happy to share perspectives on ancient people, knowledge and practices. 'They were no different from us, just with a deeper understanding of the natural world, and the importance of people and their wellbeing', Mara observes, powering up her laptop. The buzz in the room decreases as the first guest speaker enters the seminar room. Surprisingly, she is a young attractive woman who would not look out of place in a corporate boardroom.

'Welcome to country, and welcome to this inaugural short course on Australian Paleo Astronomy. My name is Melanie Campbell, and I work for the Northern Land Council. We consult with traditional landowners and other Aboriginal peoples with an interest in the way land is used and managed in the Territory. Landowners must give informed consent before any action is taken to affect their lands and seas. I live in Darwin, I am a Larrakia woman, I went to school, and grew up here in the Top End, before studying at ANU, the Australian National University in Canberra, for those of you who are not from Australia. We were really quite surprised when we got applications from people overseas. So a special welcome to our friends from the United Kingdom, Norway, and Canada. My job is Liaison Officer with the Northern Land Council, and I've been asked to give you some background information on the indigenous people in this region.

I'll leave the introductions to the first guest lecturer, Ron Jagamurra. So let me give you a short introduction to local indigenous people and our culture today, and how it got to where it is now.

There are many historical and cultural reasons why white education systems in Australia have not always accommodated the First Australians. After all, new genetic evidence indicates we've been here, the latest estimate is over 60,000 years. One of the primary reasons that we are still trying to establish our rights is the legacy of colonialism and the forced removal of indigenous children from their families, known as the Stolen Generations. Around seven years ago we collectively delivered a statement of what we want, in view of the last two hundred years. It is known as the Uluru Statement of the Heart, calling for establishment of a First Nations Voice enshrined in the Constitution, and a process of truth-telling between our people and the governments.

In recent years, there has been a growing recognition of the need to decolonise education and to prioritise indigenous knowledge systems and ways of knowing. There have been efforts to incorporate indigenous perspectives into the curriculum, to support the development of indigenous-led education initiatives, and to address systemic barriers to access and success for indigenous students. There is still a lot of work to be done to ensure that we have equal access to quality education that is respectful of our cultures and values.'

The Northern Territory is renowned for its rock art sites, which showcase the artistic traditions of Indigenous peoples over twenty thousand years, according to the archaeology. The rock art galleries feature hundreds of intricate and vibrant paintings, drawings, and engravings depicting various subjects such as people, animals, spirits, and Dreamtime stories. The field trips you'll be going on over the next two weeks are visits to rock art sites including Burrungkuy which you may have heard of as Nourlangie, and Madjedbebe, the oldest known site showing the presence of humans in Mirarr Country in Kakadu National Park. And you'll have the opportunity to look at the stars at night, and learn how ancient people viewed the constellations under southern skies.

I hope this short course on indigenous Australian paleo astronomy will give you a taste of the rich tradition of stargazing practised by our people over the millennia since we first found our way, by land and sea, to this country. When sea levels decreased, due to glaciation, Australia and New Guinea formed a single landmass connected by a land bridge across the Arafura Sea off the coast from Darwin, with only a short sea voyage required to cross the Torres Strait.'

Melanie pauses, to survey the attentive faces of the participants seated in the semi circular arrangement in front of her. 'Oh, in case you are wondering why I have a Scottish surname, the basic version is that while a small number of surnames probably come from romantic relationships and marriages, many of our surnames have their origins in the administrative requirements by government and police for identification. Often these names were selected for indigenous people or sometimes mothers chose their own. Many were chosen based on who owned or worked at local pastoral properties.

I hope you enjoy this course. Now let me introduce you to Ron Jagamurra, Indigenous Astronomy lecturer from Melbourne University. Today he'll be telling you a little about the traditional star lore of the Anangu people of the Central Desert culture.'

'Hi all, good morning, my name is Ron, and I am a proud Warramangu man from Tanami, north of Alice Springs, in the heart of the Central Desert.

In the next two weeks, we'll be providing an introduction to traditional astronomy of both the Mirarr people, and my people, the Warlpiri, and the Anangu people. We'll go on a field trip to Kakadu, Mirarr country in the second week. We'll also be looking at the use of the stars for sea navigation by the Torres Strait Islanders. We have a number of guest speakers, including a park ranger from Uluru. So let me start by giving you some background.

There are many indigenous Australian nations, comprising different clan groups and language variations. Anangu refers to related clans living across the Central Desert. The Yankunytjatjara and Pitjantjatjara people are the traditional landowners of Uluru. While there are different traditions about the Milky Way and the constellations, there is a lot in common from country to country.

For example, the Coalsack is a dark cloud just below the Southern Cross. For many Aboriginal people across Australia, this is the head of the Emu in the Sky, which stretches from the Southern Cross along the dust lanes of the Milky Way to the constellation Scorpius. Unlike non Aboriginal people,we see the dark spaces of the Milky Way, rather than the stars, as important, because this is where the Emu in the Sky is seen.

Our people are keen observers of the night sky, having detailed knowledge systems built around the Sun, Moon, and the visible planets, Mercury, Venus, Mars, Jupiter, and Saturn. For countless generations, we memorised the apparent movements of the planets and stars by observation. Oral tradition, song and dance were the methods used by our people to preserve their knowledge to pass on to successive generations. Aboriginal and Torres Strait Islander people of course distinguished planets from the background stars, and recorded the changing positions of planets in the sky over days and seasons, noting their positions relative to each other, and noticing relationships between them on their journey across the stars.

In many traditions, the planets are seen as children of the Sun and Moon, ancestor spirits walking across the sky, connecting ceremony and indigenous law to different constellations of stars. There are different cultural stories from different mobs.

The people of the Australian Central Desert practise a rich and complex system of celestial knowledge that has been passed down through generations for tens of thousands of years. Our people have developed a deep understanding of the movements of the stars and planets, and have used this knowledge to navigate through country, track seasonal changes, and importantly to inform cultural practices. Any questions so far?'

Mara ventures. 'Do indigenous Australians mark the passing of the seasons with the stars?'

Ron continues. 'Yes. One important aspect of Central Desert astronomy is the use of the night sky as a seasonal calendar. The stars and constellations are used to track the changing seasons, important for predicting the availability of food, water, and other resources. For example, the rising of particular stars or constellations can be used to predict the arrival of certain animal species, the onset of particular weather patterns, or the time to undertake certain cultural activities, such as ceremonies or the harvesting of particular plants.

Another really important aspect, particularly for Central Desert people, is the use of the stars and constellations for navigation. The stars were used to guide people across the vast and often featureless landscape of the desert, and to orient in relation to particular landmarks or features, and most importantly to find water and food sources.

Indigenous Australians developed sophisticated navigational techniques over millennia, based on the position and movement of the stars, allowing us to find water and food, and safely travel vast distances across the desert with great accuracy.'

Mara ventures. 'There are new findings and analysis of prehistoric cave art, such as the Lascaux paintings, that suggest ancient Europeans used star alignments for at least the last 20,000 years too. Some of the animal symbols in the caves may have represented star constellations in the night sky.'

'That's interesting. Are you the astronomer from Scotland? '

'Yes, Mara Stewart, I lecture at Edinburgh University.'

Glad to have you here, Mara. We have open discussions later in the week, comparing indigenous with western astronomy. Looking forward to your contribution. Are you willing to give a presentation to your fellow students? ' asks Ron.

'Of course', replies Mara.

'Great.' Ron is clearly pleased at the prospect. 'Everyone, this is a great opportunity to engage and learn more, and see how European constellations compare with what we developed here. While it differs from one aboriginal nation to another, I can tell you that the Yolngu people of Arnhem Land understood that the Sun and Moon move in an east to west motion, the Moon goes through repeated phases that affect the ocean tides. They also knew that the Earth is finite in space, and the Moon covers the Sun during a solar eclipse. I am not sure whether they understood that the moon's orbit of Earth wobbles, gradually causing the moon to rise at different points on the horizon over the years till it returns to a former rising and setting point once every eighteen and a half years. They may well have done. Like indigenous hunter gatherer people everywhere, we were very observant while still living a traditional lifestyle. We had to be, because we depended on country, the land, the animals, the plants and the seasons, for survival.

We also know that central desert astronomy is closely tied to our culture and spirituality. Amongst the different mobs, the stars and constellations are seen as spiritual entities, and appear in many traditional stories, songs, and art. Considered to be an important link between the physical and spiritual worlds, and appearing in many cultural practices, including ceremonies, healing rituals, and initiation rites. By the way, we use 'Mob' as a term identifying a group of Aboriginal people associated with a particular place or country. It is used to connect and identify who an indigenous person is and where they are from. Mob can represent your family group, clan group or wider community group.

This week we'll be joined by Dr. Noel Pembroke from Perth, who'll tell you a little about some of the Aboriginal dreamtime stories of creation and the stars of the Noongar people. Next week, we'll be going on an expedition to do some star gazing ourselves, out at Kakadu National Park. And of course enjoy a guided outside experience to observe the night sky. You can look through a large portable telescope and marvel at the wonders of our universe, as viewed from the Northern Territory. Let's take a break for lunch and meet back here at,' Ron consults his phone.' say one thirty.'

Mara and Grace exchange glances of keen interest as they head out to find a cafe.

CHAPTER 8

At the end of the winter term, Jonas decides to volunteer to spend a few weeks in the field. Through correspondence with friendly fellow marine biology students, he learns of a project, engaging with various aspects of reef ecosystems, including coral health, biodiversity, water quality, and climate change impacts on the Great Barrier Reef.

He is excited at the prospect of helping to collect data through field surveys, underwater monitoring dives, and remote sensing technologies to assess the reef's condition and track changes over time. 'And a chance to snorkel and dive the reef in brilliantly clear blue waters while making a difference? Why not?' he decides. He's accepted as a volunteer, assisting a team of marine biologists, during the Norwegian summer vacation. The project is reef restoration, focused on enhancing the reef's resilience, and damage limitation, in the wake of increasing numbers of coral bleaching events. He is looking forward to learning about new techniques for rehabilitation, such as coral gardening, first hand. He is determined to read all he can over the next couple of weeks before his big trip.

How did this happen, Jonas asks himself, as he sits in the library at the Drammen campus of the University of Southern Norway. He is attending lectures in this historic port city, set on a fjord adjacent to Oslo. Searching online for information about the Great Barrier Reef, he learns of the potential reef damage from bauxite shipping by mining giant Rio Tinto. Further browsing through online newspapers and periodicals, he comes across several newspaper reports of Rio Tinto, deliberately blowing up an archaeologically significant site, 50,000 years old, Juukan Gorge. As he reads through the reports, he wonders how they could be so insensitive as to think it OK to destroy a treasure trove of ancient culture. What he reads tells him that it was to profit from the Chinese appetite for Australia's iron ore, to power not only economic growth but also military power. Like his colleagues, he has followed Russia's invasion of Ukraine, alarmed at the development of the alliance between Russia and China, with the worrying global ambitions of both countries. Not that he thinks the USA is blameless.

Also like his fellow Norwegians, Jonas has not forgotten that China, with no Arctic presence, has inveigled its way onto the Arctic Council, established as an intergovernmental forum for cooperation among Arctic states and indigenous peoples, The core members of the council are the countries with a presence in the Arctic. This includes Canada, Denmark, Finland, Iceland, Norway, Russia, Sweden, and the United States, as well as six permanent participants representing indigenous peoples of the Arctic.

China's interest and involvement in the Arctic Council increased dramatically in recent years, and not always with everyone's approval. Joining European and Asian states, China gained observer status in 2013.

The melting sea ice in the Arctic has the potential to open up shorter shipping routes to access untapped natural resources. China's ambition to secure its interests in these areas, particularly in terms of trade and resource exploration, is no secret. Many see the northern route from Beijing to Murmansk as too close for comfort, militarily and culturally, especially for people living in northern Norway and Finland.

China's engagement in the Arctic allows it to participate in discussions and decision-making processes related to the future of the region, with new models predicting that the Arctic Ocean could be completely ice free by 2035. Including China as an observer on the Arctic Council, once seen as diplomacy, is now raising concern amongst the permanent members. And Norway's border with Russia is one of the internal security factors that are being deeply discussed.

Jonas rises from the table in the library building, overlooking the futuristic looking Ypsilon walking bridge gracefully spanning the Drammen Elva, the river that forms the inner part of the fjord. He walks downstairs in search of a coffee, and strolls outside to the quay, observing the boats lined up on the adjacent marina, and the sculptural concrete steps that form a dock for kayaks. He enjoys the break in the bright spring sunshine.

Once again back at his desk, his thoughts turn to the human consequences of climate change and the Great Barrier Reef. 'Greed for money from resources is like cancer', he thinks, 'because it is destroying the natural environment, particularly the ocean, which is bad, because that is the part of the carbon cycle that is the real source of climate stability.'

He flips to another article, turned up from his search, about the Ranger uranium mine, in the Northern

Territory, owned by a Rio Tinto subsidiary, Energy Resources Australia, where there are also deep concerns about the impacts of mining on local resources, particularly water. He reads on. 'The mining and processing of uranium can have significant impacts on water resources, particularly in areas with limited or sensitive water supplies.

Jonas keeps reading, as the report details concerns about the impact of uranium mining contamination of rivers and creeks in the world heritage Kakadu National Park.

He knows from the depths of his Sami roots that mining activities can disrupt natural water flows, and alter the quality of water resources, impacting aquatic ecosystems and the communities that rely on them. He is not naive, so he is sceptical about the efforts of mining companies like Rio Tinto to minimise the impacts of uranium mining on water resources.

He has deep concern about Norway's proposal to open up parts of the country's continental shelf for commercial deep sea mining in the Arctic, coming just after world leaders, including Norway, finally adopted a Global Treaty to protect the oceans. He and his friends are agreed that this cannot be allowed to happen. They are all ready to campaign against exploitation of the Arctic when the debate reaches the Norwegian Parliament, the StorTing.

'Whew. This is heavy stuff!' Jonas thinks about what he might learn from local conservationists holding the mining industry to account in the Northern Territory. He has heard of beautiful Kakadu, and like his contemporaries, would love to travel to Australia. His thinking is that he might learn something that he and his friends can apply to the campaign against Arctic deep sea mining.

He continues reading. 'The long-term damage from uranium mining at Rio Tinto's Ranger mine in the Northern Territory has a range of environmental and social impacts, including on land, water, air quality, and the health of local communities. The mining and processing of uranium can also generate large quantities of waste rock and tailings, which can pose risks to human health and the environment, if not properly managed and contained.

He stops reading, and sits back with his hands interlocked behind his head, contemplating. He decides to include Kakadu in the Northern Territory in his trip Down Under. As it's only weeks till he sets out for the Great Barrier Reef, he turns his attention back to his travel plans.

CHAPTER 9

Melanie Campbell is watching an NITV broadcast at her apartment in Darwin. 'Kakadu National Park, designated by UNESCO as a protected area in the Northern Territory, is renowned for its cultural and natural heritage. It is considered to be of outstanding value to humanity. The park is located within the Alligator Rivers Region, covering nearly 20,000 square kilometres, nearly half the size of Switzerland.

The first uranium was found in Kakadu at Coronation Hill in 1953 by a government geologist. The Commonwealth of Australia and the Northern Land Council, representing the traditional Aboriginal land owners, reached an agreement, and the Ranger Uranium Mine was established in 1980, on land owned by the Kakadu Land Trust. The site of the Ranger mine and the adjacent Jabiluka area were specifically excluded when the park was established.

Uranium ore mining ceased production in January 2021, as a result of financial losses from the market slump following the 2011 Fukushima nuclear disaster. Today the mine is owned by ERA, a subsidiary of mining company Rio Tinto, infamous for its destruction of Juukan Gorge in the Hamersley Range Pilbara region of Western Australia. The gorge held a cave that was the only inland site in Australia to show signs of continuous human occupation over 50,000 years. It was deliberately destroyed by Rio Tinto in May 2020 to dig up a seam of iron ore.'

Melanie presses pause, and goes to the kitchen to

pour a glass of white wine. It is a 2021 Hell Freezes Over Chardonnay from the Adelaide Hills. She doesn't know a lot about wine, but she knows she likes the dry, refreshing taste of this one. She couldn't resist putting it into the online shopping basket when she read the blurb - 'This is what happens when an old mate rings you out of the blue and tells you he owns one of the best vineyards in the Adelaide Hills, and would I take a look – 30 year old vines of the Gin Gin clone on a steep westerly slope based on sandy loam over sandstone at the superior Ashton end of Piccadilly Valley. After years of saying 'We'll make a Chardonnay when Hell Freezes Over'... well here you have it and it's delicious'. Sinking back into the lounge, she takes an appreciative sip.

She is well aware of the controversy surrounding decommissioning and rehabilitation of the damage to the Jabiru site, where water and waste management issues are still not being seriously addressed. Even with an agreement in place to transform the land to a condition 'consistent' with the surrounding Kakadu National Park, it is a huge mining rehabilitation challenge to fill the gigantic holes, knock down all the buildings, dispose of wastewater and tailings dams. Then there is the requirement to replace millions of trees. She wonders how that is going to work, when there are already so many invasive species. Dense stands of gamba grass and rubber vines smother and overgrow native vegetation, reducing habitat quality for native animals and disrupting ecosystems. The infamous cane toad was introduced in an attempt to control sugarcane pests, but they soon spread over vast areas of Australia, including Kakadu, bringing their toxins with them.

She presses play, and the broadcast continues. 'Rio

Tinto already has a track record for rehabilitation of the environment, not only all over the developing world, but also not far away at Rum Jungle, south-west of Darwin. Rum Jungle was one of Australia's oldest uranium mines, originally set up in the 1950s, closed in the seventies, and never rehabilitated. Acid and metals from the mine have been leaching into parts of the nearby river systems ever since. Now the mine is to be restored at a cost of, conservatively, nearly a billion dollars, a significant proportion of which is being funded by the Australian taxpayer.

The company has consistently denied any responsibility for fixing the problems it created for shareholder profit managed from London. Rio Tinto Group is an Anglo-Australian metals and mining company founded in 1873, the world's second largest after BHP, producing iron ore, copper, diamonds, gold and uranium. '

Melanie reflects that there is never going to be an attitude of understanding towards the traditional owners of the land, because the mining company's only motivation is profit.

'Indigenous Australians shaped the landscape for more than 50,000 years, harnessing and managing the ecology, to maintain biodiversity. Indigenous people have management rights to more than three quarters of the land in northern Australia. In many areas land managers generally have maternal links to land while landowners often have paternal links. Their motivation is the maintenance of land, culture and environment.

Indigenous men and women are equal before traditional law and customs, with acknowledged differences of responsibilities. All indigenous landowners and managers have specific and complementary rights and obligations to ensure the spiritual and physical health of their own traditional country, language and culture.

Aboriginal Australians hold the view that people do not own land, their responsibility is to care for their country, with an obligation to effect positive social, environmental and cultural outcomes for the generations to come. Indigenous ranger groups are now combining the latest science and technology advances with intricate traditional knowledge and cultural practices.

However in a political climate of permissiveness for mining companies, indigenous communities continue to face significant challenges. Even though rights to land ownership and control are well established, governments continue to find ways to avoid consultation. In the view of local indigenous people, the Mirarr, Kakadu is threatened by intrusive mining practices with little understanding of traditional land and environment management.'

'And', Melanie thinks to herself as the program ends, 'loss of connection to country and customary law, passed from generation to generation. And no understanding of the sacred trust involving people, ancestors and descendants, to pass on traditional stories.'

She can't help sighing as she contemplates the complex problems of Kakadu. Governments trying to usurp traditional land rights, with scant respect for country, sacred sites, or advice from indigenous park rangers. Historically, governments shamelessly conspired with the secret myths of middle class Australia to claim racial superiority. Indigenous soldiers from both world wars were treated shamefully. While recent governments have taken a more enlightened view, the damage has been done, and unconscious bias remains.

Indeed, prior to colonisation, Australia had been a pluralistic, egalitarian society with vast areas of land managed in perpetuity by traditional practices of maintaining grasslands, native forests, water resources and keeping the country fertile and productive, in harmony with native species, plants and animals.

Firestick farming was a national practice at the beginning of the nineteenth century, when soldier and convict settlement commenced. Land grabs, genocide and torture subjugated the indigenous population, and commenced the long and inevitable decline of resource management. Inappropriate European farming practices caused widespread water shortages, deforestation, and salination of the water table by the start of the twenty first century. Unsustainable farming is exacerbated by the onset of anthropogenic climate change, leading to massive bushfires, and major flooding events, accelerating year by year.

The stony outcrops and the sweeping views over the flood plains of Kakadu may be world famous, but for Melanie, returning to Kakadu means addressing the problems. A delay for the rollout of the government promised funding to upgrade Kakadu National Park is just the beginning. There is the headache of Jabiru, the Ranger mining town that the government wants to paint as a utopian indigenous tourist destination, managed and run on top of the unaddressed water and land degradation by Rio Tinto subsidiary, ERA. Another of the big issues is damaged roads, and under-resourced tourist infrastructure. So many people come to see the expressive rock art, continuous over thousands of years, as well as the beautiful expansive landscapes.

CHAPTER 10

Name: Alex Petersen, Age: 28, Occupation: Public Servant. This is the opening gambit in Alex's CV, which he has been touting around, looking for a new role to advance his career.

He grew up in Darwin and always wanted to move south to a bigger city. He pursued his education at the ANU, the Australian National University, eventually obtaining a degree in Public Administration from this prestigious institution. Driven by a desire to establish a business and to make his own money in as short a time as possible, Alex decides his best option is to stay in Canberra after his studies, to gain experience working in the beating heart of the nation's public administration.

Alex currently serves as a policy analyst in the government department responsible for environmental sustainability, Parks Australia. His role involves researching and analysing various environmental issues, developing policies, and providing recommendations to higher-level officials. He works in a team that collaborates with stakeholders from various sectors, including industry, and the UN, as well as other government agencies. His job includes liaison and negotiation with indigenous traditional owners, to address pressing environmental challenges in the country's most stunning natural Aboriginal heritage areas. Coming from Darwin, he went to school with indigenous kids, so he is comfortable communicating with traditional owners, where other colleagues are not. He quickly carves out a niche, and is made responsible for reporting on Kakadu National Park, south of his home town.

Alex's workplace is a heritage listed government office building located within the Parliamentary Triangle in Canberra. He enjoys adopting a professional attitude in his dealings with his colleagues. As a public servant, he is required to learn and maintain the service's code of conduct and ethics, designed to ensure departmental transparency and impartiality. Parks Australia is part of the federal portfolio of Climate Change, Energy, the Environment and Water. Alex's direct report is to branch head, Sandra Murphy.

He is tasked to provide evidence-based environment management reports. Traditional knowledge has to be taken into account, equally with science and research. Alex's job is to analyse particular situations that may arise, to establish strategic progressive management of the Kakadu National Park. However the reality of Australia's natural environment is that there are many pressures from resources companies that overlap with Parks mission to enhance Australia's social and economic well-being through environmental sustainability. It is not easy to harmonise management issues from a scientific and technical perspective, while there are subtle pressures to accommodate commercial activities, as well as respecting the cultural views of local indigenous people. Kakadu National Park is leased to the Director of National Parks. Cooperative arrangements between Parks and Traditional Owners, the Mirarr people, provide for joint management of legal matters as well as day-to-day operations.

From the indigenous people's perspective, Kakadu is a living cultural landscape. It has been home for millennia, part of the oldest living culture on earth. Some people live in Kakadu's towns and others in more remote parts of the park, all having a deep spiritual connection to country. While they have a lot in common, there are big differences in Park's and the Mirarr people's objectives.

Alex works hard, and does his best, aware that his contributions have a direct impact on the environment and the wellbeing of the traditional owners of Kakadu. The official aim is to preserve natural resources, combat climate change, and protect the country's unique ecosystems and biodiversity. Alex likes the thought that he is playing a crucial role as a responsible global citizen, even though he is well aware that the mining industry is exercising undue political influence, and there is an unwritten rule that the minister has to be kept happy.

For someone from the Top End, Canberra offers a unique lifestyle that combines urban amenities with a well preserved natural environment, centred around a man made lake. Since student days, Alex has enjoyed exploring the city's vibrant cultural scene, art exhibitions, concerts, and theatre. Being close to the mountains, he occasionally embarks on weekend hiking trips during spring and autumn, and skiing adventures during winter months. In summer, like many Canberrans, he heads to the beach, enjoying the relaxed lifestyle of the south coast as often as he cares to make the two hour drive each way.

He also enjoys hitting the clubs in the city he

now calls home, and socialises with different groups that he encounters, always diverse and interesting, because of Canberra's role as a national capital. He meets people from different countries, political groups, students, lobbyists, and occasionally, embassy staff. He is personable, and coming from the far north, has an engaging, open manner that allows him to mix socially and professionally with ease.

Overall, like many a dedicated young person who has chosen a career in Canberra in the federal Public Service, Alex takes pride in his work and his contribution to government. He is happy with his choices, personally, socially and professionally. He managed to find an apartment before the housing crisis, and feels that the seven years spent in the public service have put him on a reasonably good career path, with his current role in Parks providing a reasonable amount of job satisfaction.

Recently, however, he has worked out that there is no prospect for quick advancement, and that he is destined to hang about at middle management level for another decade, before he has a chance of serious promotion. Like many of his generation, he would prefer an important job title over more money. With his impulsive personality, he is not content to sit and wait for someone to retire. He has discovered that there is little political advancement available in Parks, nor is there much opportunity to transfer his skills to another department. So he decides to try to use the knowledge he has gained to look for a role in the business world. Preferably as an executive in a dynamic company with opportunities for international expansion and travel, shares and bonuses.

When he receives a summons to her office from Sandra Murphy, he wonders what it's all about, as he knocks on the door of her office, enters and takes the proffered seat on the other side of a large desk. 'Can you attend a diplomatic do? It's a cultural exchange event that the Malaysian embassy is putting on. They want a closer liaison with Parks, to see whether we can share information about restoration of degraded mining sites, given the changing climate. I don't want to go alone, and the invitation is for two people from Parks. It should be reasonably interesting. Do you have formal attire? Have you ever been to a diplomatic event? It's the usual - fostering international relations, promoting cultural exchange, and strengthening diplomatic ties between countries.' Sandra pauses for breath, not expecting a response. Alex obliges by assuming an interested facial expression. She continues. 'The email talks about the objective of promoting 'meaningful dialogue with a view to exchange of information on environment management after a mine closure' whatever that means. With the cleanup of the Ranger Uranium Mine in Kakadu, it should be right up your street.'

Alex takes in the stream of consciousness information coming from his boss, and learns that the guests will be fairly low key, as these events go, mainly mid ranking government officials, including guests from the New Guinea and Indonesian embassies, as well as Australian public servants. He and Sandra work out that it may be due to rumoured new environmental directives coming from the UN, more readily addressed with collaboration and cooperation, in view of changing weather conditions in the region. Weather events caused by fluctuating ocean temperatures in the Pacific, coupled with changes in the atmosphere, used to occur relatively infrequently. In recent years, El Nino droughts and La Nina extreme rainfall are now occurring every couple of years, affecting post mining site restoration.

'The fact that Parks Australia has been invited, as well as the other environment related agencies, makes me think that Malaysia may have come by some kind of advance notice of forthcoming changes to international regulations ' adds Sandra. 'All that is required is that you turn up, make polite conversation, and put forward the idea that Parks is interested in cooperation, without being specific.'

'OK, I can do that.' Alex feels he is a past master at making interesting noises on being cooperative without committing himself to anything, during his time in the public service.

Sandra continues. 'It is a reasonably prestigious affair, it's not being held at the Malaysian embassy's official residence . It's at the Hyatt. So fairly high-end, but not important enough to merit the embassy.' Alex agrees with Sandra's prognosis that the event will be carefully staged to reflect the status of the occasion, and the importance of the commitment to environmental management.

'A dark business suit will be fine.' Sandra continues. 'I suggest we meet in the lobby beforehand. I'll be there at seven thirty, so make sure you are there before then, waiting for me.'

'OK Sandra, I'll be there.' Alex returns to his desk, wondering whether his best suit will do, or if he should splash out on a new suit in anticipation of more diplomatic events to come in a rising career.

Turning up the following Wednesday, the atmosphere at the party is warm, cordial, and sophisticated. Alex enjoys the sense of formality and the polite conversation, as guests engage in networking. The mood is generally convivial, and it is clear from the low buzz of conversation that the focus is on building relationships and strengthening cooperation between environmental agencies and the Malaysian government. Alex notes the cameo appearance of the Australian Climate Change minister, probably to show support for international responses to changing weather.

Malaysian cultural elements are on display, playing a significant role showcasing the rich heritage and traditions. Videos run continuously, showing vignettes of the environment, history and culture of the people, back projected onto the walls, with the obvious objective of fostering relations and promoting appreciation of traditional Malay customs and practices.

Alex compliments his senior colleague on her smart, tailored ensemble. He is not sorry that he visited an upmarket department store, feeling satisfied with the suitability of his appearance, enjoying the feel of the neat tailored lines of his new suit that cost more than he wanted to pay. At least he knows he can now meet the dress code for a diplomatic party, typically semi-formal for a hotel affair. Waiters and waitresses circulate, offering wine. Tables are loaded with a diverse selection of dishes. Indeed, the food reflects the culinary richness of the country's typical fare. Nasi Lemak, considered Malaysia's national dish, is served in ceramic bowls, fragrant coconut rice with sambal, fried anchovies, roasted peanuts, cucumber slices, and boiled eggs prove popular. Rendang, satay and chicken rice, cooked in chicken broth, and a sour and spicy fish stew all make an appearance on the buffet table. Growing up in Darwin, Alex is no stranger to the cuisines of South East Asia, featuring in restaurants and on market stalls.

Sandra having joined senior Government colleagues, he is free to enjoy himself. He looks around at the crowd, immersing himself in the warmth and laughter, as people mill around the buffet, conversing while sampling the food. The communal nature of the meal encourages conviviality, and Alex soon finds himself in conversation with a group of young embassy staff members. He knows how to be engaging and interesting, and he is skillfully drawn into the conversation. It's his turn to introduce himself. 'I'm Alex Petersen, and I'm an environment sustainability policy analyst for Parks Australia. I cover Kakadu National Park.'

'Hi Alex, nice to meet you. I am Safira Pratama. I do ecotourism and green initiatives. Media liaison with the Indonesian Embassy. Here to promote good relations in sustainability.'

'That sounds really interesting, Safira, so what have you been promoting lately?' Alex wants to keep talking to this attractive young woman.

'Well, we want to lead by example, implementing

eco-friendly practices at the embassy, such as reducing waste, energy conservation, and promoting sustainable transport. I am part of a global outreach team liaising with other countries. Here, of course it's Australia, but I also talk to people with similar interests in embassies in friendly countries here in Canberra, to identify programs where we can share information' Safira is happy and clearly well primed to respond to his question. 'And I am here at the invitation of my friend Zhi Chen Lim, the Malaysian Embassy Environment Liaison Officer. He is doing a similar job to me.'

'Hi Alex, really pleased to meet you.' is the reply from the smiling and personable young man. 'And I also liaise with the UN on sustainable development.'

'Oh, good to meet you too. Am I right in calling you Zhi Chen? Do you mean the UNDP, UNEP and UNFCCC on environment, development and climate change? That is part of my job too.'

'Then we have a lot in common. Just call me Zhi. Safira and I have met a number of times at embassy events.'

Just then, Sandra catches Alex's eye on her way out, waving a quick farewell.

'Is your colleague from Parks Australia too?' Safira asks, well trained to the liaison part of her job.

'Yes, Sandra Murphy. She's a branch head at Parks Australia, and my current boss.'

They remain chatting on topics as diverse as the next UN climate change conference, and the exorbitant rents in inner city Canberra. A half an hour later, they discuss a move on to one of the clubs, at the suggestion of one of Zhi's embassy colleagues. 'Would you like to join us, Safira? We usually head to our favourite nightlife venue, Club22. It is a decent place, located in the Sydney Building, but not too trendy. They play a variety of music from every decade. Alex?'

'Yes, count me in,' he makes himself sound enthusiastic, while thinking to himself 'Middle of the road, making it a suitable destination for a post embassy party.'

In fact the choice is at the recommendation of Zhi's colleague. 'Absolutely lovely atmosphere, great music and vibe. Bartenders friendly, and the drinks are great. My favourite bar whenever I go for a night out!'

'I have been to Club22 before, and they do try to please all music tastes. Quieter than some of the other venues.' Alex responds to the suggestion. 'And they have a really well stocked bar, I can get my favourite Scotch whisky there, Glenfiddich. You can't get that everywhere.'

'Well the service is good, and the DJ takes requests.' Zhi offers. 'And I really like the unique retro style of decor. '

'That settles it then.' Alex turns to Safira. 'You'll join us? We can take a taxi between the four of us, and have a nightcap. Does that suit your homeward journey, Safira?'

'In fact it suits me really well, Alex. My apartment is a short walk from the Sydney Building.'

The suggestion adopted, Alex shares a taxi with Safira, Zhi Chen and his embassy colleague for the brief ride across the Molonglo River from the Hyatt Hotel, to the north side and the location of the city of Canberra. The iconic Melbourne and Sydney buildings precinct was built in the early stages of Canberra's formation, right in the heart of the city centre. They alight at the club, mount the stairs, and find a table. Zhi takes orders and heads for the bar, leaving Alex to chat with Safira. It is still only 10 pm, early for a night club, and the noise level has not yet started to rise.

On Zhi's return from the bar, after commenting on the club, the drinks and Canberra, the conversation turns to leisure activities. Alex explains that he often goes to the beach at weekends, with a new found passion for surfing. His companions express an interest in learning to surf, and one thing leading to another, the offer is made to make it a joint expedition the coming Sunday, weather permitting. Safira is enthusiastic, as is Zhi. Alex wonders whether there is a frisson of attraction between Safira and Zhi. He hopes not, as he wants to find out more about Safira himself.

'Oh I have always wanted to learn to surf', enthuses Safira.

'Is it hard?' Zhi asks his question. ' I am not sure I'll be able to do it.'

'Not if you get a soft beginner's board and go to a group lesson.' replies Alex. 'I've only been surfing for a couple of months. I come from Darwin, in the Northern Territory. Too many crocodiles in the water there, and in any case, no surf. There's a beginners class at 10 am, so we'll have to leave early.'

They arrange to meet the following Sunday, Alex to pick up his new friends in the city, for the drive to the south coast, and a surfing lesson at Broulee Beach.

They say their goodbyes at the door of the club, and as he makes his way home, Alex wonders whether it was really a good idea to go out of his way to invite them to the beach. He'll be stuck with their company for the whole day. However he decides that this is part of career development, a private public relations exercise. He'll forge a couple of new international relationships, and also have company for the long drive to and from the coast, which can be tedious alone, particularly considering that the traffic is bound to be intense if the weather is good.

Sunday dawns fine, and Alex drives to the rendezvous point at Civic, glad to see both Safira and Zhi already at the bus stop waiting for him. It's early enough and there is little traffic. Pulling into the kerb, he greets his new friends while they pile into his car.

'Beach here we come.' Alex is nothing if not a charming host able to put people at their ease, from years of practice. Particularly people from other countries. They head out of town, and soon swap Canberra for sheep and cattle farming country, while Alex keeps up a running commentary on the expansion of white Australian settlement inland, the post convict squattocracy, and the birth of the nation's capital, at the time of Federation. His guests are polite, and Alex is engaging, consequently soon everyone is actually enjoying themselves, as they drive through the varying landscapes on their way to the south coast.

Passing through Braidwood, with its historic buildings and rural charm, they drive onwards through hilly pastoral country. Suddenly they are amongst the cool, temperate rainforest of Monga National Park, on a mountainous road that winds as it leads down to the coast. They are surrounded by a landscape containing remnant plants from Gondwana, the 500 million year old supercontinent that existed before Africa, Eurasia, and the Americas land masses drifted away northwards.

With a couple of surfing stories to entertain his guests, the time passes quickly. They bypass Batemans Bay, heading to the beautiful, long sweeping beach that is protected at one end by Broulee Island, surrounded by state forest. Right on time, they pull into the beach carpark, parking near the local surf school trailer, holding racks of boards.

It's a sunny morning and they are warmly greeted by two women surf instructors, who introduce themselves. 'You've come back again, Alex. Good to see.'

Introductions over, Alex, Safira and Zhi change into swimsuits, and wetsuits, joining the others on the beach for a safety briefing. They learn about ocean awareness, reading surf conditions, and the importance of following safety instructions. Alex heads out, leaving the newbies being introduced to their boards on the sand, where they spend the first fifteen minutes learning the proper way to paddle and stand up on the board. Before hitting the water, the class participates in a group stretching session. As the instructors explain, this helps to prepare the body for action, before being immersed in relatively cold water, and their first experience of facing the waves head on.

With the basics covered, the beginners are coached into paddling, catching a wave, standing up, and riding it towards the shore. Alex is only just out of beginners class, so he is close enough to observe the joy on his new acquaintances' faces, as one after the other, they eventually learn to stand up, and ride their first wave. From the whoops, and shouts of joy, Alex is happy to see that it is as just much fun for his companions, as it was for him a couple of months previously.

After an exhilarating session in the water, everyone returns to the beach, carrying their boards back to the trailer, and divesting themselves of wetsuits, changing back into beach gear.

'That was so much fun.' Zhi tells Alex as he drops his board back on the grass. ' I am so glad we came here.'

'Did you enjoy yourself, Safira?'

'It was amazing. I am so hooked.'

After drying off, they head for a beach cafe for coffees and brunch, feeling the physical exhilaration, mental clarity, connection with nature, and social bonding that all surfers experience. The feeling that is responsible for so many people looking for any opportunity to return to the waves.

'What a fantastic experience. It was really fun. I stood up three times.' Safira is enthusiastic.

'That is the best. When you are on the wave it is like nothing else.' Zhi responds.

'Glad you liked it.' Alex is pleased to share his newfound love for surfing. 'I go at least every second weekend while the weather lasts. Glad to have your company.'

And by arrangement, Safira and Zhi continue to share

Alex's surfing trips to the coast over the next couple of months. Safira and Zhi are careful to pay for petrol, and they both refuse other invitations from Alex to socialise in Canberra. Nonetheless there is a real camaraderie amongst them, and they share a lot of stories on the long drives to and from the Coast.

Alex is interested when Safira tells them about her Aunt Dewi, a hotelier who started out with a small hotel in Bali, expanding throughout Indonesian coastal tourist resorts. Particularly when Safira tells them about a potential joint venture with Persekitaran, a Malaysian company which proposes to redevelop the old town camp of Jabiru into a tourist attraction, now the Ranger Mine is closed.

Alex wants more information. 'So how is that supposed to work? The development has to go to an Australian company. And be approved by both the Kakadu Board and the Australian Government. Parks Australia would have to be involved too.'

'I am not sure of the details, Aunt Dewi was talking about an Australian partner based in Darwin. She doesn't know if the deal is going ahead or not. She was told that the Darwin partner has got the approval of the NT Government. I don't know any more than that.' Safira is clearly not comfortable talking about a potential tourist resort in the Northern Territory, and quickly changes the subject. 'I am going to try out that Malibu board today that the school is selling off. It seems such a good deal.'

'End of the season deal,' agrees Alex, nonetheless thoughtful about what he has just heard.

'I love the pink frangipanis.' Safira enthuses. 'My favourite colour.'

' A Barbie board,' teases Zhi.

The now familiar Broulee Beach appears to their left. 'Surfs up,' says Alex casting an eye over the waves rolling into the shore.

He says nothing about what he knows of Jabiru. He is well aware that the Northern Territory Government is supporting the development and implementation of the Jabiru Masterplan. He also knows that a senior Northern Territory public servant had a trip to Disney World paid for by a private company seeking government funds to build a tourist information facility, and that promised commercial projects have not materialised. He has heard that successive federal governments have not completely closed off the possibility of further uranium mining.

He's read about the transition from mining to a tourism-based town, with a range of possibilities, including a crocodile free lake to swim in, a new town centre, a five-star lodge, glamping and a World Heritage Interpretive Centre. His role in Parks Australia, however, has taught him that visitor numbers have declined with the rundown of infrastructure, and the rise of invasive species in Kakadu. Surveys indicate that tourists are there for the beauty of the landscape and the natural wonders, not fancy tourist centres and luxury accommodation. The view inside Parks is cynicism about government objectives. Media articles continue to appear, claiming that Rio Tinto and ERA will continue to resist paying up the promised restoration funds. The view is that NT Government commitments to provide essential services long term for local people, such as electricity, water, sewerage, education and health services, could easily be backtracked.

Alex suspects that money and grandiose ambition will play a very big part in what happens at Jabiru. So he wants to know more about what is being secretly discussed for the post mining development of Jabiru. And from his colleagues, he has heard that Persekitaran has strong links to China.

CHAPTER 11

Back home in Skjervoy, at the end of his project, Jonas decides to fly straight to Australia.

He wakes up early, and walks to the ferry along the marina, silently saying farewell to the fishing boats, the yacht from France, proudly flying the tricolore, nestling amongst all the other boats in the harbour, and the glorious views north over snow capped mountains rising out of the dark blue water of the fjords. He looks up at the sky which has an ominous tinge, with a bank of clouds hanging over the line of hills above the town, and for the first time in months notices that instead of a calm flat water inlet, there are little waves and whitecaps stretching out into the waters beyond the ferry terminal.

Embarking down the gangplank, he shows the ticket on his phone, and settles himself into an empty row of comfortable seats, near the plate glass windows that provide a panoramic view out past the islands to the expanse of the long stretch through Norwegian waters to the remote island of Svalbard. A few minutes later, and the calm floating-on-air pace of the hurtigruten fast boat, suddenly becomes a bucking bronco as the swells roll in from the north, whipped up to heights that Jonas was not expecting, even though he is a sailor himself.

'A fitting Arctic farewell,' he muses, as he settles in to enjoy the frenzied waves crashing against the hull as the boat passes by islands, and straits between shorelines, stopping only occasionally to pick up intrepid passengers braving the elements. The two hour journey to Tromsø, the largest town in Troms and Finnmark county, passes quickly. On his way to Australia.

Arriving at the quay, he makes his way by local bus to the airport, in good time to start the epic journey by air to Queensland. Excited about the prospect of the adventures ahead, Jonas makes his way patiently through the security queue, and heads for the departure gate. From Tromsø to Oslo to Singapore is quite a hike, and he is glad of a stopover in Singapore en route to Brisbane. He has never been outside Europe, so he is looking forward to experiencing Asia for the first time. Finally, after what seems like a long weary flight, he is looking out of the porthole, not at clouds, but at the sea, and ships that look like toys from the low altitude of the A380 on the final leg of its descent. He is excitedly looking down on the Singapore Strait, a narrow water passage that separates the Indonesian island of Sumatra to the west and the Malaysian peninsula to the east.

Connecting the Strait of Malacca to the South China Sea, it is one of the busiest shipping lanes in the world. For over a millennium it has served as a crucial trade route, connecting major ports to South East Asia, and beyond. The Strait hosts numerous cargo vessels, tankers, and other commercial ships passing through it daily. Jonas observes keenly as the sea gives way to tropical vegetation and the aircraft makes its final approach, touching down at Changi airport with the roar of rubber on bitumen, rumbling, then rolling at a sedate pace to the stand. The usual disembarkation scramble means that it is another half hour before he is out of the airport. Despite the queues, the fingerprint identification system at the border is speedy enough, and having picked up his backpack from the baggage carousel, he emerges landside. He finds the marshalled taxi queue, and having stowed his luggage, thankfully slides into the back, giving the driver the name of his Orchard Road hotel.

After a sleep and good breakfast, Jonas is ready to explore. His first port of call, after reading the literature in the hotel lobby, is Sentosa Island, an Asian style beach resort, 'a popular tourist destination offering a variety of attractions and activities. Tanjong Beach is known for its picturesque setting and range of amenities. He decides to navigate the public transport system, and with the help of a map, takes a bus from Lucky Plaza, walking the last stretch to Tanjong Beach. While his preference is for more secluded and natural beach settings, given that he is in one of the most densely populated regions on the planet, he appreciates the beautiful coastal setting with sandy shores and clear blue waters. He enjoys his swim, and finds a spot on the sand to relax and soak up the sun on this glorious day, with a few puffy white clouds, and brilliant sunshine out of clear blue skies. Families are happily installed around umbrellas, engaging in a range of activities and water sports, including swimming, beach volleyball and paddleboarding. Family-friendly owing to its calm waters and lifeguard services, there are plenty of children splashing about, with parents who are clearly relaxed, stretching out under umbrellas, while the kids do their own thing.

Jonas has also picked up the literature for the Maritime Experiential Museum at his hotel. 'The museum showcases the maritime history of Southeast Asia. Featuring various exhibits, interactive displays, and multimedia presentations, explore different aspects of maritime history, trade routes, and seafaring traditions in the region' reads the brochure. He has an endless fascination with the sea, boats, and the amazing sea voyages of the past, when seafarers had simple craft with sails. Without too much difficulty, he finds the island bus to take him there from the beach. Wandering slowly through the various exhibits, he soaks up maritime history, trade routes, and seafaring traditions. He is fascinated by the deep and ancient seagoing history of this part of the world, as old as the maritime history of Scandinavia, he thinks. Jonas admires the boat designs, and the exhibits of wrecks found by divers, not only ancient Chinese junks, but also Arab trading dhows under sail.

Jonas reads about the significance of the maritime Silk Route, and is impressed by the knowledge and skills on display of the historical shipbuilding, navigation, and exploration of old. Arab traders were sailing through Singapore on their way to China well over a thousand years ago. For business from the merchants of Baghdad, intrepid sailors reached China by sea, sailing through the Persian Gulf, navigating through the Indian Ocean and reaching Chinese ports such as Guangzhou, formerly known as Canton, and Quanzhou. He is swiftly falling for the allure and mystique of the far east. The Arab traders brought goods such as spices, incense, glassware, pearls, and precious stones from the Middle East and Africa to China. In exchange, they obtained highly sought-after Chinese silk, porcelain, tea, and other luxury goods.

He makes a mental note to explore more of this part of the world in future. He is impressed by the island nation with its trees and tropical vegetation growing profusely amongst the high rise buildings. Singapore is now one of the world's busiest trading hubs and a major global logistics centre and a global financial hub including the newly emergent fintech companies.

Having checked out of the hotel before heading to the beach, when he returns, he heads to the outdoor pool surrounded by lush tropical vegetation, including huge frangipani trees, to spend a couple of hours swimming and lazing, before making his way to the airport. Browsing through a bookstore on the way back, he picked up a magazine containing an article about Singapore. Lying on a recliner in the shade, after a refreshing dip, Jonas reads that the country's robust regulatory framework, political stability, and wealth have contributed to its reputation, however there is a price to pay, with a government that always gets elected, and a quiescent population who sometimes feel they are living in a goldfish bowl.

He reads of the spread of Covid out from the crowded accommodation for 'guest workers' who collectively keep the economic engine running. He discovers that Singapore receives low-wage workers from many countries with the most significant proportion coming from Bangladesh, India, and China, in addition to Sri Lanka, Thailand, the Philippines, and Vietnam.

'Singapore's economic development since the

1960s has been rapid,' he reads, ' transforming the country from a relatively undeveloped nation, with the three main ethnic groups, Malays, Chinese and Indians, engaging in family market businesses, to a highly developed international market economy. The government actively attracts foreign investment, especially in manufacturing, by offering incentives, establishing industrial estates, and providing a business-friendly tax environment. To its credit, the government has established research institutes, promoted collaborations between industry and academia with generous R&D incentives. This focus on innovation helped Singapore move up the value chain and become a hub for technology and knowledge intensive industries. Singapore has placed increasing emphasis on sustainable development and green initiatives. The country has implemented various measures to enhance energy efficiency, reduce carbon emissions, and promote sustainable urban planning.'

Every country has its hidden secrets, he reflects, putting down the article entitled 'The Truth about Singapore's Economic Boom'. He wonders whether the people of Singapore are happy with the power structure, given that people at the bottom of the social scale are often the foundation of the wealth of the upper echelons. He changes from his swimming gear to comfortable clothes for the next leg of his journey, and makes his way back to Changi Airport. Once through to departure, he fires up his laptop in one of the many comfortable lounge seating areas, to write his impressions of his first experience of Asia, a taster for returning some time in the future.

After another night on an aircraft, he gazes out of the window at the seemingly endless flat scrubby country unrolling below him, while dawn lights up the horizon with pale yellow clouds. The landscape gradually changes to a green coastal plain, and while it is still early, they land at Brisbane Airport. The arrivals process is more or less what Jonas expects. Finding his way to the right queue for non Australian passengers, he proceeds to the immigration area. Finally it's his turn, and he presents his passport, having completed his arrival card. He has the necessary eVisa that Norwegian citizens require. With his visitor pass, he can stay in the country for up to 90 days, long enough for his volunteer project.

The official on the desk asks him the purpose of his visit, and he provides a description of the reef research project. 'Good on you. Enjoy your stay, Mr. Jansen. The reef needs all the help it can get.' With that, Jonas is granted entry into Australia. He collects his luggage, having declared he is not bringing any foods, plants, animals, or other restricted items into the country. However he is selected for a baggage inspection. He decides it's not random, it's because he is a young foreign male. He is directed to the inspection area where his belongings are screened by a large electronic machine. Finally through the formalities, he blinks at the brightness of a sunny Brisbane morning, as he exits the arrivals area, emerging into the large light filled terminal building. Australia, at last. He looks around for transport options, spots the signs for taxis, rental cars, and public transport. After some judicious questions at the information desk, he decides on a train into

Brisbane en route to the north.

Believing in spur of the moment travel plans, he decides to hop on a northbound bus the next day. First to the coastal town of Seventeen Seventy, because he is fascinated by the name. Then to continue north. He has, after all, four weeks before he is scheduled to turn up at his volunteer project with the Australian Institute of Marine Science in Townsville. He is looking forward to the experience, to support coral reef resilience, collecting data on species biodiversity, the core of any healthy, functioning reef. If there is a high level of diversity, then there is also a thriving population of all other marine life. Seagrass meadows are one of the most important reef habitats, providing food, shelter and nurseries for many marine animals. In Queensland, they are critical feeding grounds for the largest grazers – the turtle and the dugong – which rely on seagrass meadows to survive. Jonas is excited about observing and learning new techniques and skills to apply to conservation in the fjords of Norway, where the pristine meadows are home to corals of genetic variation, hopefully resilient enough to cope with climate changes in the fast changing Arctic region.

Alighting at Roma Street, he finds the Brisbane City Backpackers, within walking distance, surrounded by parklands, convenient for an evening stroll. The staff are friendly, the four bed dorm turns out to be decent for the price, and he settles himself for a good night's sleep, in anticipation of an early start for tomorrow's first leg of the journey north. He has a thirty day travel pass that allows him to hop on and hop off long distance buses across Australia. He is starting with the ten hour journey to Agnes Waters and Seventeen Seventy. Beating the busy Brisbane morning traffic with an early departure, the bus soon escapes from the suburban sprawl, and the eucalyptus tree country unfolds on both sides of the highway.

It is truly a beautiful scenic journey through some of south east Queensland's stunning landscapes. The bus offers comfortable seating, clean toilets and good wifi, meaning the trip passes pleasantly. It is hot, so Jonas is thankful that there is on board air conditioning. He enjoys the journey, as they pass through a diverse range of coastal vegetation, and magnificent tree stands surrounded by pasture for grazing. They motor along, passing tropical plantations of pineapples, macadamia, mangoes and avocados, interspersed with native Queensland bushland. The driver keeps up a running commentary, in true tour guide style. 'Our trip today passes through fascinating destinations like Gympie and Maryborough, two towns rich with history, character and heritage architecture. We'll pass by Hervey Bay, one of the best places in the world to go whale watching. It is known as the gateway to the spectacular beaches, breathtaking rainforests and the unique ecosystem of world heritage listed Fraser Island, where you can swim with sea turtles, kayak the crystal clear creeks, climb the dunes and watch the many species of seabirds, eagles, ospreys and many more. Further along the way we'll pass through Bundaberg, home of the world famous Bundaberg Rum Distillery. Next time you pass by, don't forget to go to the Turtle Centre, dedicated to marine turtle research, protection and education, to see the largest concentration of nesting Loggerhead turtles in the South Pacific.'

The day passes quickly, and Jonas is glad when the bus finally arrives at Agnes Water in the late afternoon. He shoulders his pack, and after asking for directions, finds the backpackers at Seventeen Seventy, the next beach along. He is amused to learn that the official town name is the words, not the numbers, so named after the visit of James Cook and his ship, the Endeavour, in May 1770. The beach, Cook's second landfall in Australia, lies between the Seventeen Seventy Point, and Bustard Bay, a seascape that gently wraps around towards the north and the port city of Gladstone. Finding the hostel, he is greeted by a large verandah with tropical plants, comfortable outdoor living space, overlooking a garden replete with Balinese style open huts. Checking in, he is really glad to find a spacious dorm room with a touch of style. Resisting the urge to head for the bar, and some friendly company, he resolves to get a good night's sleep, then spend the next couple of days exploring the area. He quickly falls into a deep sleep, still exhausted from the long journey from the Arctic Circle in Northern Norway, to Brisbane in Queensland, on his way to the deep tropical north.

The next day, he wakes early in the morning, listening to the unique cacophony of sounds made by the birds. 'The raucous squawks of Black Cockatoos eating their way through the she-oak seeds. Rainbow Bee Eaters, Drongos, Butcher birds, Friar birds, Black faced Cuckoo shrikes, Curlews, Kookaburras, together with Honey Eaters and Finches are all present in our beautiful local environment.' Jonas is fascinated by the names of the birds that he reads from the leaflet in the reception area of the hostel.

Looking out of the window from the breakfast room, he spots a pair of kangaroos looking in his direction. 'Or are they wallabies?' he wonders. He gathers his day pack, swimming togs, sunscreen, water, hat and towel, as prescribed by the eco tour company providing a day trip to the Great Barrier Reef that he signed up for the previous evening through the front desk. 'Join an unmissable Queensland experience on a reef cruise out to Lady Musgrave Island! Day Tour departing 1770 at 8 am. Lady Musgrave Island is an ocean built coral cay located in the southern Great Barrier Reef. We drop anchor in the crystal clear turquoise lagoon with its colourful healthy reef gardens. Share the raw beauty of this natural wonder through the glass bottom boat, guided island walk, swimming and snorkelling.'

Jonas makes his way down the road to the point of departure, buying a coffee and a breakfast roll at the marina cafe, before joining the queue of tourists from diverse backgrounds, ages, and ethnicity, waiting for the arrival of the boat captain and crew of the catamaran hydrofoil. Once on board, he settles himself on the front deck, as the boat gathers speed passing the Seventeen Seventy lookout, heading off the coast under a bright blue sky on a hot sunny day.

After a safety briefing, the passengers position themselves in the cabin and out on the deck to take advantage of the stunning scenery. The man standing beside him observes 'No wonder they say that the weather in Queensland is beautiful one day, perfect the next.'

Jonas smiles. 'My second day in the country.'

'Where are you from?'

'Norway. From Oslo. Jonas, Jonas Jansen.'

'Welcome to Australia, Jonas. My name is Graham, Graham Saunders.'

'Good to meet you, Graham. Have you been in Australia long?'

Graham smiles. ' You could say a long time. I am an indigenous Australian. What brings you to this country?'

'I am volunteering to assist in a reef restoration project with Queensland University. I am a marine biologist.'

'That's great. The reef needs all the help it can get.'

'Funny, that's what the man at the immigration desk said at the airport.'

'Australia was in climate change denial for too long, so we are starting behind the eight ball. Also farming and mining have played a significant role in weakening the environment, as well as global warming. Sadly, Australia has the highest per capita carbon footprint in the world. Not something to be proud of.'

Jonas warms to the softly spoken man standing beside him on the crowded front deck, and they settle into a companionable silence as the boat slices its way through the swell. After a few minutes, they are treated to some tourist information over the loudspeakers, by one of the tour guides.

'The boat is specifically designed for reef excursions and is equipped with amenities to ensure your comfort during the journey. The crew members, who are knowledgeable about the reef and its marine life, welcome you warmly. Now we'll provide an overview of the day's activities, and then a safety briefing.

Enjoy this short four hour scenic journey through the sparkling waters of the Coral Sea. The Great Barrier Reef has significance as a World Heritage site. Lady Musgrave is the second southernmost island in the Great Barrier Reef chain. The island is located on the southern end of a large lagoon,which we enter via a deep water channel.

During the boat trip, you can enjoy the stunning coastal views of Queensland's coastline. The lush greenery of the islands and the pristine beaches add to the natural beauty surrounding you. If we're lucky, we might even spot some marine wildlife like dolphins or turtles swimming alongside the boat.

As the journey unfolds, many people choose to remain on deck on such a beautiful day, rather than sit inside the comfortable cabin even with the panoramic windows. Jonas and Graham amongst them, chat desultorily, about the reef, the weather, the climate and the beautiful seascape around them. As they approach the island, the boat slows down, making its way through a narrow channel to anchor in the lagoon, and the crew prepares snorkelling gear and wet suits, while welcoming the arrival of the glass bottom boat that has joined them from the direction of the island.

The tour guide is speaking. 'First we'll have morning tea, then we have two main activities. We provide you with high-quality snorkelling gear and offer instructions on how to use it safely. If you're a beginner, don't worry, the crew is experienced in assisting all skill levels. And we'll take it in turns to take the glass bottomed boat to walk around the island.'

Jonas really enjoys his experience of the bright blue underwater world teeming with life. The coral reefs, with dazzling colours and intricate formations, create a mesmerising sight. Through the glass he can see an abundance of tropical fish of all shapes and sizes, moving effortlessly among the coral, that seems to him to be very healthy. He spots fish, rays, and even a small reef shark. In response to his question, the tour guide replies 'this far south, the reef is still in really good condition.'

The walk on the island is memorable, and the crew members are knowledgeable, pointing out birds, tree species and interesting features as they walk across the island, coming back via the beach. The group is receptive to anecdotes about wildlife, island history and safety, and after further exploration of underwater wonders, everyone returns to the boat for a well-deserved break. After a lunch of tropical fruit and sandwiches, people return to the deck, finding spots out of the blazing sun, to relax while taking in the breathtaking views and the intense turquoise of the water. As the afternoon draws on, they start the return journey to Seventeen Seventy. People share stories with fellow travellers about their underwater adventures, and Jonas and Graham pick up their conversation from where they left off earlier in the day. Jonas explains to Graham that the reef is one of the world's most magnificent natural wonders, who smiles at this revelation from a Norwegian to an indigenous man. All too soon, as the sun starts to drop below the skyline, the coastal hinterland comes into view as they get closer to Seventeen Seventy.

'Captain James Cook, captain of His Majesty's ship The Endeavour, came ashore and landed on the beach of Round Hill Creek. He wrote in his journal that he saw the top of the peaked mountain to the southward above the clouds. Thus the name. Later renamed Seventeen Seventy.' Graham observed.

'I don't suppose your people are grateful, as you lost your land.'

Graham maintains a diplomatic silence. 'So are you going directly to your project after this?' he asks, as they wend their way through the sand bars on their approach back to the beach.

'No. I have four weeks before I am due there. So I am planning to head north, then travel across to the Northern Territory. I have an open 30 day bus ticket to go anywhere. I can please myself, and I want to see the real Australia'

'Good for you. But I think it might take more than four weeks. You can only scratch the surface. Australian civilization is at least 50,000 years old. Maybe older. They are not sure, though new DNA evidence points to somewhere around 70,000 years of continuous occupation. If you make it to Darwin, be sure to visit Kakadu National Park. It is very special.'

'It's on my list.' The boat docks back at the marina, and the passengers disembark, somewhat wearily, after a long day of sun and sea. 'See you, Graham, nice to meet you.'

'You too, Jonas, take care, stay safe'.

Jonas walks the short distance back to the backpackers, picking up a pizza from the cafe. Once again, he is in bed early, and falls into a deep sleep almost as soon as his head touches the pillow. The next morning he makes a decision to see more of the country before he returns to Queensland to join the research project. He heads for the middle of town, and picks up another bus to take him to Rockhampton, for a leisurely journey up the coast to Townsville. From there, he takes a long hop all the way to Darwin in the Northern Territory.

CHAPTER 12

Graham Saunders flies to Sydney from Melbourne's Tullamarine Airport. He has an important broadcast to attend. He's done his research. The coronation of a new king in England is a significant ceremonial event that symbolises the official ascension and investiture of a monarch to the throne. Upon the death or abdication of a monarch, the event is officially announced, and preparations for the coronation of the new monarch begins. The coronation typically takes place in Westminster Abbey, a historic church in London that has been the traditional venue for English coronations since 1066. The abbey is transformed for the occasion, with elaborate decorations and seating arrangements. The new king usually travels in a grand procession from a royal residence to Westminster Abbey. The route is lined with crowds of well-wishers and spectators eager to witness the historic event.

Upon reaching Westminster Abbey, the new king is received by clergy, nobles, and officials. The ceremony commences with various rituals, prayers, and readings from religious texts, emphasising the spiritual significance of the coronation. The new king takes a solemn oath, affirming their commitment to uphold the laws, protect the Church, and govern the realm with justice and fairness. This is followed by the anointing with holy oil, symbolising the sacred nature of their role.

Henry the Eighth split from the Roman Catholic Church and created the Church of England because Pope

Clement VII refused to grant him an annulment of his marriage to Katherine of Aragon, so that he could wed Anne Boleyn. Henry named himself Supreme Head of the Church of England. And subsequently beheaded Anne. His daughter, Mary Tudor, reversed all of this when she came to power in 1553. She restored Roman Catholicism as the state religion and the Pope as the head of the church. Her half sister Elizabeth, succeeded to the throne, later beheading her cousin, Mary Stuart, Queen of Scots, in 1587, before restoring the Church of England.

Graham Saunders was invited as a guest commentator on the ABC prequel to the coronation of Charles the Third, not as a journalist but as an indigenous man. Men and women were arrested as late as the 1940s when policemen overheard them calling to their children. It was forbidden to use Aboriginal languages in public then. In fact the history of Australia includes many massacres in the name of the English crown after the military invasion of Australia, bringing with them, the poor convicted men and women from filthy, disease-ridden prisons of England. There was no treaty, no peaceful settlement, and a false claim that Australia was terra nullius. A pattern of brutal reprisals and genocidal killings began to emerge in the late 19th and 20th centuries as thousands of Aboriginal people were murdered especially in the NT and WA. During what has become known as the 'Frontier Wars', Aboriginal and Torres Strait Islander nations and tribes fought to defend themselves and their country, with many violently murdered in the clashes with white settlers. New research into the scale of the violence during that time suggests the massacres of Aboriginal people became larger, more organised and ruthless as the decades went on.

Graham politely questions the relevance of the crown to Australia. At the time of Australia's invasion by the British, a Germanic George was on the throne, sadly suffering from mental illness.

The pinnacle of any coronation is the crowning of the new king. The Archbishop of Canterbury places the crown on the monarch's head, signifying their authority and sovereignty. The king also receives other regalia, such as the sceptre and orb, which represent various aspects of royal power. After being crowned, the new king is hailed by the congregation and members of the nobility, who swear allegiance and pay homage to the monarch. This is a symbolic act of recognising the king's authority and pledging loyalty. The Australian Prime Minister, in attendance, swore allegiance to this man, whose distant ancestors included petty officials from northern Germany.

Graham walks away from this experience with the national broadcaster, the ABC, after coverage of the king's coronation prompts a torrent of vitriol towards him from Australians who probably do not believe in Father Christmas or the Tooth Fairy, but who clearly do believe in the Royal Family.

He decides to follow up on a story as a freelance correspondent that he heard last time he was up north.

Persekitaran is a privately-owned conglomerate, and its ownership is primarily attributed to the Chang family. The company was founded by a prominent Malaysian businessman of Chinese descent, who has played a significant role in its development and expansion. Sri Tan Tang has been the key figure behind Persekitaran's growth and success. However, as a privately-held company, specific details about the ownership structure or distribution of shares within Persekitaran are not widely disclosed to the public, though it is widely believed that the Chang family continues to maintain control of the company.

It has diverse business interests, including timber, palm oil, media, and property development. While Persekitaran is primarily associated with Malaysia, it has extensive operations in Southeast Asia and the Pacific. In recent years it has acquired business interests in Australia in various sectors, primarily timber and forestry targeting the international market. It has established timber processing facilities in Australia, Brazil and Indonesia. Not without controversy.

Graham is following his journalistic instinct. Since the debacle with the ABC coronation program, he has been following up on leads of promising stories. Reports of endemic corruption of local government, particularly in Queensland, lead him to Townsville and a rumour about a gated tourist development. He reads a Townsville press release 'Persekitaran flatly denies any secret deals between an Australian company with links to the Townsville Council. Persekitaran has a genuine background in tourism, and is deeply committed to benefiting the local community. And there is absolutely no evidence that Persekitaran's extensive portfolio of businesses has degraded the land with which it is associated. On the contrary, it has provided beneficial influences to local communities in many other countries. '

But Graham has found some interesting reports, linked to related claims of illegal logging, tax evasion, and various human rights abuses, all covered up through subsidiaries and shell companies disguising the company's real plans and activities. As yet he has not found any hard evidence In fact, even the initial plans for a project entitled 'Ocean Blue' north of Townsville refer to a gated community for business executives, not a new closed tourist resort for Chinese visitors, which is what he is hearing on the grapevine. Graham's sources tell him that the tourism business plan is rumoured to be a front for secret development of a much more sinister nature, to be rubber stamped by local officials through local companies.

The secrecy surrounding the purpose of the Persekitaran development is only matched by the reluctance of a senior local government official to answer any questions about the Ocean Blue project approval. The same official refuses Graham's invitation to meet, to explain and answer questions.

'If there is nothing to hide', a fellow journalist seated at the bar informs Graham, 'then all the involved parties would have contacted the local papers, consulted the Townsville Council long ago, and been fully transparent and accountable. They would have sought publicity for the project activities. And yet there is the undisputed fact that there was an illegal demolition of houses inhabited by army veterans, ostensibly by a local contractor's shoddy work. The same contractor insists that he was paid by the developers, who told him they had the necessary permissions. This resulted in denial of responsibility by Ocean Blue, and the bankruptcy of the contractor, and extensive asbestos contamination of public beaches and environmentally fragile mud flats nearby, all in the very central area proposed for the Ocean Blue project development. Put to rights by the ratepayers of Townsville. Dirty deeds done dirt cheap indeed.'

Graham finishes his drink. 'Thanks for the intel. There is no smoke without fire, I'll see what else I can dig up from the archives,' he remarks as he takes his farewell, and makes his way along the Esplanade back to his hotel.

He has had a tip off from another journalist that Persekitaran is behind the Australian company that wants to turn Jabiru into an international holiday destination post the clean up of the Ranger Uranium Mine. For whom, he can't help wondering. Can his source be right? Can the plan really be to turn Jabiru in the Kakadu National Park into an international gated holiday experience destination for the mega wealthy? What happened to the much touted glamping resort? Have they approached the Northern Territory Government? He can't help wondering if there is some more sinister purpose behind this business, given Persekitaran's reputation. He books a ticket to Darwin, and leaves early the next day.

CHAPTER 13

Mara and Grace meet up, by arrangement, for a coffee at the campus cafe before the start of lectures. Noticing another course participant in the queue, Grace issues an invitation to the young man with untidy blonde hair who is also picking up a coffee. 'We introduced ourselves yesterday on the course. I can't remember your name, but I remember you are from Norway. Would you like to join us? I'm Grace, by the way, and this is Mara. She is a proper astronomer from Scotland. I'm from Sydney. It is great to see so many non Australians on the course.'

'Jonas. And yes, I would like to join you.'

Grace and Mara explain their interest in astronomy, and then Grace ventures. 'So why are you here? What brings you down under? Are you an astronomer too, like Mara?' Jonas laughs. 'No. I happened on this course by accident. When I arrived here a couple of days ago, I went to the student accommodation service to find a place to stay for a week. I was planning to just look around, you know, go to Kakadu, the Mindil markets, all the tourist things. Then I saw this leaflet on the notice board outside the student accommodation service. As I have two weeks to spare, I decided this is an opportunity not to be missed. I get to hear the real stories of aboriginal people, and to go on a field trip to Kakadu. And meet new people. '

Coffees paid for, they find a table by a window that looks out on a courtyard where a large poinciana, also known as a flame tree, is already showing its spectacular display of bright red and orange flowers. 'I really like the tropical vegetation here' observes Mara, as they sit companionably enjoying cappuccinos for Jonas and Grace, and an americano for Mara. 'So different from Scotland, where the landscape has been largely deforested and farmed for over five thousand years.'

Grace, as the one Australian at the table, elaborates. 'They are all over the parks and gardens in Darwin. I love the deep red flowers. They first appear around this time of year, and stay on the tree for a month or more. And they have a typically tropical wide canopy, which we don't get in Sydney. Look over there across the lawn. It's a paperbark. They also grow abundantly here. They're found in the wetlands and coastal areas. You can see their peeling bark, which looks like paper, hence the name. They often grow near water. See down there, in the hollow, you can just make out the reeds growing around the creek at the end of the park. There are loads of them flowing down to Darwin harbour.'

'You are very knowledgeable about plants, Grace?' asks Jonas.

'Not really, I just like to look at them. Australia has a gorgeous range of tropical and subtropical vegetation. You can't help being impressed by the trees, flowering shrubs and bushes. Especially here in the north, with the abundance of water in the wet season.'

'When is the wet season then?' asks Mara. It has been lovely sunshine since I arrived five days ago.'

'October to April' replies Grace. 'May to July are the months that experience relatively cooler weather. The temperatures range from around 17 to 23°C'.

Mara and Jonas exchange glances. 'And this is cool weather? In Europe, that is high summer, even as far north as Scotland.'

Jonas adds 'In Norway, this is a really good summer.'

'In Scotland average temperatures are increasing across all the seasons. Summers are now warmer and we get less rain. While this is challenging for the environment and farmers, we do get to go to the beach more often! After all, it is an ill wind that blows nobody any good.'

Jonas and Grace both laugh. 'Except that climate change is so serious, it's hard not to worry about it. And here in Australia, the Great Barrier Reef is in big trouble. After this course is over, I am joining a research project on the reef off the coast of Townsville, collecting data on the health of inshore coral, seagrass and water quality.'

'Brilliant. That's really good.' The women chime in response.

Mara asks. 'So how did you get to land such an interesting job?'

'It's not a job. I am a volunteer, through a collaboration with my university. I am doing postgraduate studies in the health of seagrass off the coast of northern Norway. So, just a lucky chance that I grabbed with both hands. Climate change is really affecting us, because the Arctic is melting so fast. Some climate scientists are saying that the Arctic has warmed nearly four times faster than the rest of the planet since the seventies.'

Grace sounds surprised. 'I didn't know it was as fast as that. Here we have serious problems with bushfires and floods that are becoming increasingly intense and more frequent.'

'Yes it is incredibly serious ' Mara provides her opinion. 'In Scotland it's warmer and wetter in winter, and the summers are hotter and drier. We even get heat waves now. And no-one is doing enough in the UK, or in Europe for that matter. Particularly because it is the developing countries like sub Saharan Africa, and BanglaDesh that have the biggest problems either from droughts and water shortage, or from coastal inundation and floods. Although people don't seem to care, that is starting to change in the past year or two. Even the Americans are starting to get it, with the frequency and intensity of fires and storms. Well we'd better head off to the seminar. We don't want to miss the lecture about Anangu astronomy and Uluru.'

They make their way across the courtyard, and having enjoyed one another's company, finding three adjacent seats they sit beside each other by tacit agreement.

Ron takes the floor.

'Overall, the traditional astronomy of the Northern Territory and the Australian Central Desert people has rich and complex systems of knowledge, playing an important role in culture, spirituality, and way of life just as much today as it did for the past tens of thousands of years. We'll also be touching on the people further south, the Anangu. Uluru, the famous big red rock, is a sacred site for the Anangu people. They have lived in the area for tens of thousands of years and have a rich cultural history associated with Uluru.

As promised, today we'll be hearing about Anangu cultural beliefs. We'll be hearing from a Uluru ranger. She's going to tell you about the famous Australian icon, the rock itself. I can tell you that they have a complex spiritual belief system that centres around the concept of Tjukurpa, which translates to 'Dreaming' or 'Creation'. According to Anangu, Tjukurpa is the time when ancestral beings created the world and all living things. Many of the stories and beliefs associated with Tjukurpa are connected to specific sites in the landscape. This is Bernadette Duncan, a ranger from Uluru, to tell you more about her people, culture and astronomy. '

A young woman, smartly dressed in a ranger's dark green uniform, with the red, yellow and black aboriginal flag emblazoned on the sleeves , stands in front of the class.

'Hi everyone. I come from the western desert culture. We are Yankunytjatjara and Pitjantjatjara people, the traditional landowners of Uluru-Kata Tjuta National Park. We speak our own language and teach it to our children. In our language we call ourselves Anangu. It is pronounced arn-ung-oo, and we ask you to use this word for our people too.

For us, Uluru is a sacred site that is associated with many important Tjukurpa stories and rituals. We believe that the rock itself was formed during the Dreamtime by ancestral beings, and that it has spiritual power and significance. We believe that the rock is home to numerous spirits and that it has a strong connection to the land and the environment. Our people have lived in the area around Uluru for tens of thousands of years, and we have a deep knowledge of the local environment and its resources. All across the western desert, people have a rich cultural heritage that includes hunting and gathering, storytelling, art, and music.

In recent centuries, the arrival of European settlers has had a significant impact on the Anangu people and our way of life. Many of our traditional lands were taken over by Europeans, and our culture and spirituality were deliberately suppressed. However, we have continued to fight for our rights and for recognition of our cultural heritage. In 1985, Uluru was returned to the Anangu people, and we now manage it jointly with the Australian government.

Anangu traditional astronomy and practices have been around for a very long time. The night sky has played a significant role in our cultural traditions and practices for tens of thousands of years. Our star stories are closely tied to spiritual beliefs. The stars and constellations are associated with specific ancestral beings and stories from the Dreaming. We believe that these stories are passed down through generations and are an important part of our cultural heritage.

We also use the stars to navigate and to find our way through the desert country. One of the most important features for us is the Milky Way, which is known as Kanyini. We believe that Kanyini is a sacred pathway that connects the earth and the sky, and that it is associated with the creation of the world and the ancestral beings.

Our astronomy is a complex and very nuanced system that reflects the deep connection between the people and the natural world. It is an important part of our cultural heritage and is still practised and celebrated today. Any questions so far?'

There follows a lively discussion amongst the course participants about Uluru, the numbers of visitors, and the reasons for respecting the wishes of the Anangu people not to climb the rock, which turns out to have practical as well as spiritual implications.

'Uluru has been sacred to our people for tens of thousands of years, and climbing the rock was not generally permitted under our law. The climb can be dangerous and more than thirty people have died while attempting to climb Uluru – many others have been injured. So we welcome visitors, and if any of you come to the rock, I'd be happy to take you on a tour.' Talk over, the students spontaneously applaud her, clearly appreciative. 'Thanks everyone, I'll hand you back to Ron now.' Bernadette smiles, as she heads out into the corridor.

Ron moves back to the centre of the room. 'We have our first field trip next week. The first day we'll head for Kakadu National Park. We'll stop to see the mysterious walls of the Nourlangie rock art site, a shelter and art canvas for twenty thousand years. There you'll be able to see Creation Ancestor Namondjok, and Creation Ancestor Namarrkon, the Lightning Man. Then we'll drive on to have lunch at the Maguk Campground, where we can observe the night sky. Maguk is located within a monsoon forest, with an easy walk which leads to a waterfall and the base of a beautiful plunge pool. The water is crystal clear and swimming is allowed. In the afternoon we'll drive to a campground, where we'll spend the night. The following morning, we'll spend some time exploring, then we can take a short hike to Boulder Creek Falls, a Kakadu favourite trail through lush monsoon forest. We have three days, driving from site to site, camping and stargazing, before we return via the highway to Darwin.'

'Can we go up to the Gunlom Falls and swim in the infinity pool?' asks one participant.

'No. The Gunlom walking track to the lookout and pools above the falls is closed at the request of the Traditional Owners. There is a dispute with Parks Australia over a sacred site. Now I've whet your appetite for the activities of next week, let's find out about your level of astronomical knowledge. We can go round the room and provide some background information on your experience, and if you like you can share your motivations for being on the course before we break for lunch. '

Mara, Grace and Jonas decide to have lunch together in the cafeteria of the campus building. As they settle into their seats at a hard-to-find single table, exchanging comments about the next week's planned expedition above the hubbub of chatter. With salads and iced teas in front of them, the conversation turns to course fieldwork, post colonial Australia and the amazing people who are still in residence, the oldest continuous civilisation on the planet.

'I had concerns that there may be too much vague theory and not enough input from indigenous experts, but I am so pleased to be wrong. This course is deeply rooted in indigenous knowledge and culture, and I am keen to learn all I can.' Mara remarks.

Grace offers her own insights and ideas as a third generation Australian, descendant of late nineteenth century Scottish immigrants, whose journeys to the other side of the world were a direct corollary of the infamous Highland Clearances, when renegade clan members facilitated the English moneyed classes to take the highlands and the glens, forcing the people from their homes to replace them with sheep.

They talk through the possibilities of next week's three day field trip around the Kakadu National Park, for some night sky observations and star storytelling led by local elders, before returning to Darwin, and the closing session of the course. They are all very excited at the prospect of visiting sacred sites of deep significance to local people. And they are fast becoming firm friends.

'Can't wait. This is why I put up with that interminable trip halfway round the world. To see the real Australia.'

'Me too, Mara, believe it or not, most Australians have no knowledge or experience of indigenous culture, for all its millennia of history. ' Grace adds 'I am just looking forward to next week's field trip. It sounds great.

'Camping out under the stars. Natural infinity pool, stargazing with powerful telescopes. Learning from the people who are still connected to the way we all lived in the past.' Jonas is animated as they rise from the lunch table.

They head back to the seminar room for the afternoon session, all three feeling invigorated and optimistic about their exchanges, and particularly about the field trip. They take up their places around the U shaped seating, with the large internet connected screen the focus at the front of the room, ready to tackle the next session.

'As it's Friday' Mara ventures, at lunch at the end of the week. 'Why don't we go out to a pub. We could have a drink and eat.' At the close of the afternoon session, Mara, Grace and Jonas discuss meeting for dinner. ' There are several places with good seafood. I would like a view over the harbour.' Grace says, looking up from perusal of Maps on her laptop. 'The Jetty, Crustaceans on the Wharf, The Oyster Bar, and Fiddler's Green. There's loads. They are the ones I can see that are close to the water.'

'Fiddler's Green' says Jonas ' sounds Irish.'

'Right first time. Their website says 'Fiddlers Green Irish pub is situated in the magnificent Darwin waterfront precinct with ocean views, fun activities and an overall fantastic location & ambience. Designed with an 'old-style Irish pub' theme, Fiddlers Green makes you feel right at home through our warm & nostalgic atmosphere, friendly staff & extensive menus."

They arrange a time to meet. A couple of hours later, Jonas walks into the pub. The atmosphere is lively and as authentic as an Irish pub can get, located in the heart of a bustling port city of Darwin in Australia's Northern Territory. Stepping inside to the warm glow of dim lighting, he hears the sound of traditional music filling the air, and a buzz of voices coming from the bar. The walls are adorned with classic memorabilia and Irish pub decor, creating a cosy ambiance. He looks around, and finds Mara and Grace already seated at a table by the windows. He walks over. 'Been here long?'

'No, just got here. We went round the shops after we left the University, then made our way here. Walked in, looked around, and decided this looks like a good table with a view of the harbour if you squint. So good timing.'

They study the drinks menu on the table - 'The pub offers an impressive selection of Irish beers on tap, including Guinness, Smithwick's, and Kilkenny, along with a variety of Irish whiskeys and spirits. The friendly bartenders are skilled at pouring the perfect pint of Guinness, and they may even show off some traditional Irish flair while doing it.'

'Have you tried Guinness, Jonas?' asks Mara, after a moment's study.

'No never. It's an Irish beer? Guinness it is'.

'Grace?'

'Same here. A middy if they have it.'

' A middy?'

'Oh that's Sydney speak. Depending on where you are in Australia, you can get a middy, a pot, schooner or a pint. The hotter it is, the smaller the beer size, because unless you're swilling it down it's too warm by the end of the glass.'

'I'll go. And that makes it Guinness all round then.' Mara heads for the bar.

The hubbub at the bar does not skip a beat, with a babble of noisy conversation making itself heard over a local live music group performing a set. Tonight, a talented group of musicians takes the stage, playing lively traditional jigs, reels, and folk songs. The infectious music has the crowd tapping their feet and clapping along. Mara pays for the beers, and carries them to the table. Conversation is limited, as the music picks up pace. They watch as a couple of the pub goers take to the dance floor in an open area near the stage. They clearly have traditional Irish dance skills, and they showcase impressive footwork, while others look on, clapping and stamping encouragement.

As immediately as the music starts, it stops. The thoughts of the three turn to food. They move into the quieter dining area in the next room, picking up menus on the way. They choose from a range of seafood, including barramundi, snapper and crab. Perfect complements to Guinness,' comments Grace. They silently flick through the choice of classic favourites like hearty Irish stew, fish and chips, steak and seasonal vegetables, bangers and mash, reading that the pub boasts using fresh and locally sourced ingredients. They make their choices, and provide their selections to the hovering waiter. The conversation turns to next week's expedition.

'Safe to say, I've never been on an indigenous astronomy field trip in a remote part of Australia before' observes Mara, sitting back as they finish the meal.

'Sure to be a fascinating experience' agrees Grace. 'Has to be a unique perspective on the night sky and the culture of the indigenous people of Kakadu.'

'I'm hoping to meet some local people, ' Jonas adds his perspective. 'I come from an indigenous community myself. Have you heard of the Sami people in Northern Norway? We have a rich cultural connection with the stars, planets, and celestial events too.'

'Really interesting. I've read a little about Sami astronomy, ' replies Mara.' I am ashamed to say that as an astronomer, I really don't know much at all about our European traditional knowledge. So I am really looking forward to an introduction to the cosmology of the local people, who after all have been doing it for tens of thousands of years.'

'I am hoping we'll be welcomed in the Aboriginal way, with a traditional smoking ceremony performed by elders. The people are Mirarr. I think I read that on the first night they do a 'Welcome to Country' , or maybe Ron mentioned it sometime during the week?' Grace provides her local knowledge. "Welcome to Country' is a formal ceremony acknowledging traditional people's land ownership. It's held at many events in Australia. '

'Looking forward to that, ' Mara is enthusiastic.

'Anyone for another beer?' asks Jonas. 'My turn to go to the bar'

'Yes please.' replies Grace.

'And for me.' echoes Mara.

Jonas heads for the bar, which by this time has more people standing elbow to elbow, vying for the attention of the two attendants serving drinks.

Back at the table, Mara and Grace continue their exchanges. 'Hopefully they'll share Dreamtime stories and creation myths associated with stars that we can actually see. The night skies are especially clear here, with no background light pollution. That will make it a really special experience. To hear about the spiritual significance of the night sky from the people who own the stories.' It's Mara who makes the comment, while they sit back and wait for the beers to arrive. 'Away from city lights, in the vast skies above the grasses and eucalyptus trees, there'll be great dark sky locations for stargazing. And learning from people who have a deep knowledge of the constellations you see here in the global south.'

'I'm looking forward to hearing the aboriginal names for the stars and constellations.' Grace responds slowly from the repleteness that comes from eating an excellent meal of fresh seafood.

Jonas returns from the bar, with someone in tow.

'Mara, Grace, this is Graham. I've invited him to join us. We went on a boat trip to the reef together. At least he was on my boat trip, and we talked all the way going out, and all the way coming back. He explained how Captain James Cook explored the coast of Australia. And how the British never asked people about taking their land.'

'Hey Graham, good to meet you. I'm Grace, and this is Mara. She's from Scotland. '

'Nice to meet you too. Jonas insisted on bringing me over to meet you. Apologies for interrupting.' Graham makes to go.

'No, please stay. Please join us,' urges Grace, pulling out the spare chair, while Mara shuffles up to make some space.

Jonas places the drinks on the table. This time there are four beers.

'Are you Australian, Graham?' asks Mara. ' Don't mind me, I'm Scottish, we are always nosy. '

'Yes.' Graham replies. ' I am definitely Australian. So what brings you all the way from Scotland, Mara?'

'Oh, that's easy. I'm here to attend an indigenous paleo astronomy short course with Mindil University. And that's where I met Grace and Jonas. We are all on the course together. We are going on a field trip next week to Kakadu. And we have the exciting opportunity to meet the local elders, who are going to tell us about their ancient knowledge of celestial navigation.

'And hopefully they'll tell us the aboriginal names for the various stars, planets, and constellations.'

'The sky is often viewed as a map of ancestral stories and relationships.' offers Graham.

'Oh you are interested in astronomy too?' asks Mara.

'Well yes, though I don't know as much about it as I would like. I know that the universe is part of aboriginal culture and dreamtimes since way back. It was important to the ancestors for survival. Gathering of food, movement of animals, movement of themselves through the seasons.'

'Strangely enough. It meant a similar thing to my Scottish forebears. So what part of Australia do you come from, Graham?'

'I live in Melbourne now, in Wurundjeri country. For us, I know that lessons of Wurundjeri Dreamtime that are written in the stars are being passed down to the next generation. But I don't know too much more than that, because I grew up in an urban environment, went to school and got a career. At that time, not much knowledge was being passed down, because it was just after the Stolen Generation debacle, when aboriginal kids were taken away from their parents and put in missions, or adopted into white families.'

'To Australia's shame' comments Grace, circling the beer mat around the table with her glass.

'The Sami people have been treated badly too' says Jonas. 'During the Nazi occupation in World War Two, my family was treated worse than the other Norwegians. We had to build shelters in the middle of the Arctic winter to survive. But at least the children stayed with their families, even though they were in deep poverty.'

'Scottish Highland Clearances were pretty brutal too,' remarks Mara. 'The new landowners replaced the people with sheep, forcing them to leave clan lands that had been home for generations and generations. They even burnt a grandmother alive when she refused to leave her cottage.'

'Why is it that some people seem to abuse the power that money brings. And more to the point, why do we let them do it?' Grace asks the rhetorical question.

They are all silent for a few moments, sipping Guinness.

'So Graham, are you going to join us on our stargazing field trip to Kakadu next week? ' Jonas asks. 'Our course is starting out at Nourlangie Rock Art, and finishing up at Kambolgie campground. '

'Tell me more' Graham is listening attentively.

'Well every night there are local elders telling us star stories while we do some stargazing. That's really all we know at this point.'

'It's been a great experience so far. ' It's Mara who continues. 'We've heard discussions about how indigenous cultures used the night sky to determine seasons and develop calendars. And how celestial events played a crucial role in guiding cultural practices, such as hunting, gathering, and ceremonial activities. There is a lot of interest in western astronomy now, as we try to determine what the people of Britain and Europe knew about the night sky prior to the Neolithic, the Stone Age. We are trying to tell whether they knew that the Earth was a sphere prior to the Greeks, and what kind of navigational techniques people were using, because not only the land, the sea was a major highway too.'

Grace adds. 'There is plenty of archaeological evidence that people were trading and travelling long distances. They've found continental pottery grave goods in the far north of Scotland in the early Bronze Age, for example.'

Mara picks up the thread. ' I am really interested in the way the stars were used for navigation at that time. This course is giving me a great perspective on how people used the sky in a practical way thousands of years ago.'

'Aboriginal people all over Australia see the night sky being an integral part of their worldview. It's different from clan group to clan group. There are about five hundred different Aboriginal peoples in Australia, each with their own language and territory. You'll have a great time on the field trip. I know that the Mirarr, the Kakadu people, have tours for cultural activities like traditional dances, bush tucker experiences - that's bush food, Jonas and Mara, and storytelling sessions around the campfire.'

'So how long have the Kakadu people been here, Graham? In Kakadu, I mean?' asks Jonas 'and where did they come from?'

'Generations of Mirarr have lived on and cared for this country for tens of thousands of years. Sharing and looking after country have always been a part of the Kakadu story.' Graham looks thoughtful, as though he is trying to mentally place the sequence of events that led to the settlement of Australia. 'As I understand it, analysis of maternal DNA shows that a significant number of Aboriginal populations moved into Australia around 50,000 years ago. Like all the human migration paths, they moved along the coastlines. As it turns out they rapidly swept around the west and east coasts at the same time, meeting around the Nullarbor just west of Adelaide, a few generations after arrival. And there may well have been earlier migrations we just don't know about at this stage.'

'Archaeology is discovering more about Australia all the time. There has not been a lot of understanding of indigenous culture in the past.' says Grace. It's only the last fifty years or so that there has been real respect for the sacred sites and traditions of indigenous communities. Now Australians are starting to understand the importance of preserving and respecting their cultural heritage.'

'So why are you in Darwin, Graham? That's me being nosy again.' asks Mara.

'Well. A few reasons. I am a freelance journalist following up on a story. It's a little complicated, but have you heard of Beetaloo Basin?'

A chorus of two nos and one yes.

'For the benefit of Jonas and Mara, Beetaloo Basin has a geology that contains what's known as 'unconventional gas resources', that is, fracking. There are well documented risks to the ground and surface water resources from mining pollution. Governments are largely ignoring these reports in favour of the miners. Beetaloo is about 500 km south of here,' Graham replies. 'And water pollution does not stay in one place, because all the aquifers are connected, and they all flow into Australia's Great Artesian Basin.'

'Oh no.' Mara's response is a drawn out expression.

'Oh yes,' echoes Graham. 'The local farmers and the traditional owners both share your view. The fear is that rivers and water sources in the region will be polluted by the waste produced by fracking. The mining industry has a terrible track record, no matter what they put in their environment management plans. They rely on cover ups, delays and obfuscations, often walking away from damage done.'

It's Mara who continues. 'So more carbon emissions, more droughts, floods and fires, as well as risks to water supply for everyone. I thought Australia already has water problems. Isn't the Murray River drying up? '

'Yes there are big problems there that are yet to be solved.' Graham continues to explain. 'There is a proposed Middle Arm industrial development here in Darwin harbour, with the federal government spending $1.5 billion of taxpayer money to be 'a key enabler' for the export of gas from the Beetaloo basin. This is the government's own words in a document released under freedom of information. And they've labelled the port development as a 'sustainable development precinct'. They deny it is a subsidy for fossil fuels. They call it 'an important way of setting up our economy for a sustainable future".

'They'd never get away with that kind of deceit where I come from.' Mara responds. 'At least not until recently, when the British government issued more oil and gas licences for exploration off the coast of Scotland. People are in so much trouble with their energy bills and cost of living, the politicians think they can get away with it. For God's sake, every ton of greenhouse gas is hastening the shutting off of the Gulf Stream.'

'Yes I read about that. ' Graham looks thoughtful. 'I am also reporting on the clean up of the Ranger Uranium Mine in Kakadu. I sometimes do interviews for the ABC, the Australian Broadcasting Corporation, and I'm in Darwin catching up with people from the Northern Land Council, and the Kakadu Board. The story is that funding for the multibillion dollar clean-up of the Ranger uranium mine is doubtful, and there are attempts to revive uranium mining. Rio Tinto is the majority shareholder in Energy Resources Australia, which owns the tapped-out Ranger mine and the neighbouring Jabiluka deposit, and by the way, the company that blew up Juukan Gorge. The Mirarr people, whose lands cover both mining leases, are adamantly opposed to any further expansion. But minority shareholders in ERA have been pushing for the reopening of Jabiluka to mine deep below Ranger. And of course ERA doesn't want to do a proper cleanup.'

'Of course. Greenpeace reports that French nuclear companies are dumping radioactive waste in Siberia, and that governments around the world are lying about safe nuclear waste disposal operations. There is a huge and growing problem of radioactive waste stored all over the world with no permanent solution. And how can anyone seriously engage in more fossil mining, when we all know that burning fossil fuels, ironically laid down in previous mass extinction events, is the primary cause of climate change.' Mara's rhetorical question echoes in the silence of the others around the table, while the fiddle and guitar combo floats in the background.

'You are clearly not Australian. We were all taught climate denial until a recent change of government, and we are now moving on to knee jerk reaction.' Grace has no illusions about public opinion on climate change in her home country. 'And we are being fed the line that the only economically responsible thing to do is to mine more coal and gas.'

'That is a great shame,' Jonas is clearly surprised at this revelation. 'And I am ashamed of what is happening in Norway. Although we know the facts about climate change, Norwegians are also doing more exploration in the North Sea. Even though we are very rich, some of us just want more money.'

'The greed factor is everywhere, it doesn't matter what country you live in.' responds Graham. 'I am also doing some background research for another story. The ABC has uncovered secret government business that should not be secret at all. An investigative Four Corners program revealed that Washington is planning to build dedicated facilities at Tindal air base, south of Darwin. The Australian people have a right to know about defence policy. The government has not been up front about it.'

'I am horrified,' Grace responds. 'I had no idea. Tell me more Graham.'

'Tindal air base dates back to WWII when it was home to a squadron of B24 bombers in defence of Australia. It was constructed by the Americans in 1942, however by the time construction was complete, the war in the Pacific was ending. The Japanese began withdrawing from the region, so it fell into disuse. Until now. Six nuclear-capable B-52 bombers are to be stationed there. The ABC found this out from US budget papers, not from the Australian government. And that's not all. A new US Air Force mission planning and operations centre is to be built, to support the aircraft here in Darwin over the next two or three years. The worry is that it locks Australia into any future military conflict between China and the US.' Graham's voice trails off.

'I am so sick of all this hush hush stuff. We have a right to know what's happening on our soil. I seem to remember there is a military base doing satellite surveillance at Pine Gap south of Uluru? I am pulling a few threads from deep in my memory banks now,' laughs Grace.

'You mean the top secret CIA base located deep in the outback that began operations in 1970 at Pine Gap, with US military intelligence gathering, and two secret rooms that no Australian is allowed to enter. Potentially making Australia a nuclear target? ' Graham prompts.

'That's the one. I'd forgotten all about it until now.' Grace looks thoughtfully into her beer. ' The price of the US coming to our aid militarily against the Japanese in the Pacific in World War II. It's true we would have been invaded without them. But still. That was a long time ago.'

The conversation turns once again to the stars and the beauty of Kakadu, its history and traditions. As the evening progresses, the music gets even livelier, and the crowd grows more exuberant. The pub stays open until late, allowing everyone to enjoy the convivial Irish spirit, no doubt until the wee hours of the morning. However at the end of the third beer, everyone decides to call it a night, and with farewells all round, they head off to their respective accommodation, Jonas and Graham exchanging numbers, before he heads back to the bar to meet a friend.

Jonas, Grace and Mara step outside the by now heated atmosphere of the pub, into the tropical evening of Darwin's winter crisp night air. There is an atmosphere of camaraderie amongst them, as they say good night and part with a feeling of amicable companionship.

All three are now sure that their respective decisions to attend an indigenous astronomy field trip in the Northern Territory, after an enjoyable evening of music, food, and good company, are fortuitous. They are all looking forward to a once in a lifetime opportunity to see the cosmos through the eyes of the original custodians of the land, gaining insights into their profound relationship with the universe and appreciating the deep connection between the night sky, culture, and spirituality. They feel like they are a part of the preservation of ancient indigenous knowledge and traditions, a deep conviction that they all share.

CHAPTER 14

Dewi Pratama is a highly educated and driven woman who comes from a successful entrepreneurial family. She obtained a degree in business management from a reputable Indonesian university and later pursued a master's in tourism and hospitality from a renowned institution, giving her the knowledge and skills required to fuel her ambition to run tourist facilities for international guests.

Over the twenty years since she started out with a small beach hotel in Bali, Dewi Pratama now owns and manages a chain of luxury resorts and hotels located in popular tourist destinations across Indonesia, including Bali, Lombok, and Komodo Island. She attracts an upmarket clientele. Her properties are known for their unique design, cultural authenticity, and commitment to sustainable tourism practices, and visitors come from all over the world.

As a businesswoman, Dewi, despite her small stature, and unprepossessing appearance, is a visionary leader who inspires her team with her passion for Indonesia's rich cultural heritage and natural beauty. The hotel workplaces are voted by her staff as very good. She fosters a culture of teamwork, innovation, and employee development, empowering them to deliver exceptional guest experiences. And this is reflected in glowing reviews from her guests.

Dewi has a global outlook and is well-versed in international business practices. She regularly attends tourism conferences, networking events, and trade shows, forming partnerships with travel agencies, airlines, and other industry stakeholders to boost tourism in Indonesia. Being environmentally conscious, she is committed to sustainability and ecotourism. Starting with her flagship hotel in Bali, her facilities incorporate green initiatives, such as renewable energy sources, waste reduction, and local community involvement in recycling waste to energy, to preserve Indonesia's natural beauty for future generations.

Dewi takes pride in promoting Indonesia's diverse cultural heritage. She collaborates with local artisans and businesses, showcasing traditional arts, crafts, and cuisines to enrich guests' experiences, and support local communities. And now she is attending events in different countries, promoting Indonesia as a tourist destination, at the behest of her government. Now her hotels are running so well, she takes up a new part time role attending functions to promote Indonesia to the world as a sustainable tourist destination, as a successful Indonesian businesswoman who has achieved prominence in the tourism industry.

She has just finished setting up a new resort at Tablolong Beach, Kupang West Timor. It is a gated tourist facility, sited on one of the most beautiful beaches on the island. It's a co-development with the Indonesian government, to be fully environmentally sustainable, with new green jobs for the locals. The site has easy access to the beach, close to cultural attractions in Kupang, and the deal includes providing local recreational activities for Indonesian tourists.

The tourist facility has been designed with a blend

of traditional Chinese architecture and Indonesian influences. The building features traditional Chinese pagoda-style roofs, intricate wooden carvings, and the vibrant red accents, symbolising prosperity and good fortune in Chinese culture. The hotel is surrounded by well-manicured gardens with lush greenery and decorative elements, creating a peaceful and inviting atmosphere for visitors.

'The China-Indonesia Friendship Tourist Center aims to provide Chinese tourists with a comfortable and familiar environment while offering the chance to explore and experience Indonesian culture,' reads the website. 'The staff at the centre are proficient in both Chinese and Indonesian, ensuring smooth communication and assistance for Chinese tourists. Please note that the overall aim is to create a welcoming environment for Chinese tourists while celebrating the cultural diversity of Indonesia.'

Dewi has been approached by a representative from Persekitaran, a Malaysian company. It was a fairly simple email, requesting a meeting with her in Bali, to discuss the possibility of a joint venture with her company to set up a similar resort in Australia. The email goes on to explain that Persekitaran are working with a local Australian company, Top End Tourism, to develop the tourism potential for the mining town of Jabiru, now that the decommissioning of the Rio Tinto Ranger mine is underway. They compliment the China-Indonesia Friendship Tourist Center in West Timor, and ask if she would be interested in a discussion about the potential project in the Northern Territory of Australia.

Dewi continues to consider this suggestion, as she

goes about daily business at her home base in Bali.

CHAPTER 15

Appearing as an online ad in the International Cruising Forum - 'Crew wanted, preferably familiar with the Asia-Australia monsoon system for Bali to Darwin trip at the end of the cyclone season, hopefully before the trade winds start. 45 foot yacht. Must be experienced sailors, preferably used to Indonesian waters reefs and shoals. Prepared for the wind direction to be against the sailing direction, if so, may have to tack much of the way, depending on wind and weather. Leaving at the beginning of May. Must be accustomed to bigger waves which may be going against the sailing direction. Sure to be a great trip. If this sounds like you, send us an email with details of your experience and availability. Leaving within a week.'

The 1500 kilometre trip starts out from Serangan Harbour on the island of Bali, catching four hours of the ebbing tide in the direction they are headed, southeast down the island chain, then to Kupang, then Darwin. The first part of the trip is motoring through channels past mangrove edged shorelines and small islands. No wind. The first evening is spent moored at the end of the channels. There is usually a six weeks window at this time of year to set out and make the trip before the trade winds set in. The worry is that they may be early.

The following morning, they set off on another low wind day. The forecast is for variable winds gusting up to 15 knots, with occasional squalls. Once again they start off motoring. Out of the channel, at last there is a bit of wind, and the mainsail goes up. Followed by a jib.

They sheet in hard to catch as much force in the sails as possible. Happy to be sailing, not motoring. Small islands are disappearing off the stern. For the present, lovely sailing conditions, next stop West Timor. Once out in the ocean the sailors on the yacht Perihelion do not see any sign of civilization. No fishing boats and no freighters on this part of the ocean. Only water until the horizon, wherever you look. The sailors pass the time either doing chores, maintenance, sailing, reading, or enjoying the spectacle of dolphins playing in the water under the bow of the boat.

Next morning early, under bright sunny skies, the Perihelion joins the numerous boats already anchored in Kupang. By now the breeze has picked up and the chop is rather intense. The boats are all bobbing at anchor. Moving further down the anchorage, they tuck in directly in front of Teddy's Bar, gaining a holding in the sand bottom for the anchor. On shore, the music is blaring and the welcome flags are flapping, an overall lively atmosphere of hustle and bustle of cars and people. They all go ashore to spend the day. The weather is holding, and they'll do the Kupang to Darwin leg starting in the evening, hopefully just over two days' travel. That night is spent dealing with occasional squalls, luckily under a full moon, meaning reasonable visibility. They've made pretty good progress overnight, taking it in turns to be awake for two hour shifts. The following morning there is a wonderful spectacle on the Timor Sea - vast expanses of luminous sea and sky, with small puffy clouds on the horizon. Nice light breeze, meaning they can sail at 6 knots, and mid morning, the mainsail goes up, allowing them to pick up speed to a steady 8 knots. Unusually lucky sailing break for this time of year, when it's often either becalmed or very windy. The weather holds, and the Global Marine Traffic AIS screen shows only a few boats in the area, as they make their way towards Darwin. Another 24 hours, and they anchor off the western tip of North Vernon Island, thirty kilometres from their destination, calm enough for everyone to get a good night's sleep. Was there a silent slapping of oars, and a distant low engine noise of a tinny? Everyone is too tired to care.

The next morning, they make it to Darwin Harbour, with a combination of sail and motor in the light winds, to the Australian boarding station for arrivals, on the northern side of the sandbar outside of Cullen Bay. They are met by the Department of Agriculture and Border Force. Officials conduct a very thorough search, utilising advanced equipment, including drug-sniffing dogs and portable X-ray scanners to detect hidden compartments and contraband. Nothing found, the crew are allowed to disembark in Australia. Greg and the crew disburse into Darwin.

Police reports of Northern Territory drug dealers organising couriers to take money to Bali for investment in clubs and bars for the purposes of money laundering, have been circulating for years, if not decades. Everyone knows it's happening. The money is alleged to be from the proceeds of drugs imported into the country. Investigators believe that as well as money being smuggled out, there are couriers going the other way, from Bali to Darwin.

The Bali Darwin drug trade has been a concern for law enforcement agencies for some time, due to its involvement in the trafficking of illegal drugs, particularly methamphetamine and MDMA. The trade takes advantage of the proximity of the two locations and the relatively porous borders of the region. The agencies have been working hard to combat drug trafficking and have undertaken various measures to address this issue, including increased border security, intelligence-sharing with Indonesian authorities, as well as efforts to raise public awareness about this dangerous, damaging, illegal trade that poses serious risks to individuals and society.

People face the death penalty if caught trying to smuggle crystal meth into the Indonesian resort island of Bali. One of the known routes is couriers who fly from Bangkok, then to Bali. From Bali, there are a number of routes into Australia. Australian law enforcement have identified that one of the routes used to avoid going through customs, past sniffer dogs at major airports, are cruises leaving from Bali, having started up again post Covid. And there are private sailing trips that cross the Indian Ocean between Indonesia and Australia with stops along the way, some heading for ports in Queensland, others for Darwin. Or perhaps a rendezvous somewhere along the remote coastline with a fishing boat.

During investigation operations, investigators have located and seized a number of drug hauls with multi-million dollar estimated street value. The NT Joint Organised Crime Task Force has accumulated a lot of intelligence on the drug trade in and out of Darwin. They now suspect illegal drugs are being brought in from West Timor and Bali on private yachts. Despite thorough inspection at the border, the task force is worried that little has been found recently.

CHAPTER 16

Mirarr country is a stunning place. Most Mirarr land is within the World Heritage listed Kakadu National Park including the vibrant wetlands of the Jabiluka billabong country and the stunning sandstone escarpment of Djidbidjidbi (Mount Brockman). The Mirarr estate extends beyond Kakadu to Western Arnhem Land as well as surrounding the Ranger uranium mine, the Jabiluka mineral lease and the mining town of Jabiru.

The relationship between the traditional owners of Kakadu and Parks Australia is complex, reflecting the long history of indigenous custodianship and the establishment of Kakadu National Park, operating under the Environment Protection and Biodiversity Conservation Act.

Kakadu Park is a vast expanse of diverse ecosystems, home to numerous plant and animal species, and rich in cultural significance for its traditional owners, who are primarily from the indigenous Mirarr speaking language groups. These people have lived in the region for tens of thousands of years, maintaining a deep spiritual connection to the land. The Aboriginal Land Rights (Northern Territory) Act 1976 recognizes the rights of the Mirarr people as legal owners and custodians over the land, which includes the right to hunt, fish, and carry out cultural practices within the park. Under traditional law, the people see things differently. They have a deep and complex system of law and governance in relation to Kakadu, which is based on their beliefs, customs, and traditions that have been passed down through generations. Their key view of Kakadu is that it is their ancestral land, with which they have a profound spiritual connection to the landscapes, plants, animals, and significant sacred sites, used for ceremonies. This spiritual connection is central to their identity, their food gathering, their song, dance and visual arts, and the way they interact every day with each other and the land itself.

The Mirarr people have a deep sense of responsibility for caring for the land and protecting its cultural and ecological values. Their law encompasses a set of traditional rules that govern various aspects of life, including social relationships, resource use, ceremonies, and ecological management. These customary practices have been developed and refined over thousands of years to ensure the sustainability and harmony of their relationship with the land. The law includes an intricate understanding of the seasonal cycles, which informs their practices such as hunting, gathering, and fire management. This knowledge is critical for maintaining the health of the land and the well-being of the community.

Their law is deeply intertwined with the kinship system, which governs relationships and obligations between family members, clans, and language groups. This system plays a significant role in decision-making, resource sharing, and resolution of conflicts within the community. The keeping of knowledge is still transmitted through oral traditions, including songlines, storytelling, and ceremonies. These cultural practices are preserved by the elders. Important knowledge about the land, its history, and the significance of specific sites within Kakadu are passed on to the upcoming generations in this way.

'Care for Country ' is a fundamental principle in everyday life. This is the responsibility of the people to protect and manage their traditional country sustainably, ensuring the well-being of the natural environment for all members of the community.

Officially, as custodians of Kakadu, they play a crucial role in collaborating with Parks Australia and other stakeholders to manage the national park, while maintaining their heritage and connection to the land. This means that at least they have some leverage over a very different set of objectives delivered from Canberra.

In 1987, Kakadu was inscribed as a UNESCO World Heritage Site. On a practical level, this makes the joint management between traditional owners and Parks Australia a working relationship. In theory, both parties share responsibility for the park's management, decision-making, and conservation efforts. In practice this is a bone of contention, as the Mirarr people's experience is well documented as being on the receiving end of some heavy handed decision making by Parks.

The Kakadu Board is a significant component of the joint management arrangement. It comprises representatives from the Mirarr, Parks Australia, and others. The role of the board is to guide the development of management plans, setting of priorities, and making decisions that align with the park's conservation objectives.

Parks Australia proclaims that it works closely with the traditional owners to protect and preserve the rich cultural heritage of Kakadu. This includes safeguarding sacred sites, rock art, and other important cultural landmarks that hold immense spiritual significance for the Mirarr people. And yet they have often failed to consult when making decisions about where to site tourist facilities. Case in point is the issue regarding Parks Australia's decision to build steps close to a sacred site near Gunlom Falls in Kakadu National Park. The situation revolved around the construction of visitor infrastructure, specifically steps leading to the top of a natural area known for its stunning natural beauty and cultural significance to the Mirarr people.

The Mirarr traditional owners expressed their unhappiness with Parks Australia's decision to build these steps abutting an area that holds such deep significant cultural and spiritual importance. Their concerns ranged from Parks' failure to consult before commencing construction, and the sensitivity to environmental impacts, affecting the local flora, fauna, and ecosystems of the Falls.

Legal action in the Northern Territory Supreme Court found the works would constitute an offence under the NT's sacred sites laws, but that Parks Australia could not be held criminally liable because of the wording of the legislation.

CHAPTER 17

'The life of an Aboriginal police officer in the Northern Territory Police Force can vary based on individual experiences and circumstances. Indigenous officers play a vital role in serving their communities and bridging the gap between law enforcement and Indigenous cultures.'

Ben is reading the brochures presented to him as available career opportunities in the Northern Territory for when he finishes high school in Brisbane. They are limited. He does not have the artistic talent of some of his peers. While he can sing in tune, there was no chance he would play in a band that becomes another Yothu Yindi. That boat has already left the dock. He did not want to spend endless years at University to become a lawyer, a doctor or a vet, even though his results were good enough to get him an affirmative place at one of the universities down south. He'd confided as much to the careers advisor who'd interviewed him in his final term. As much as he loved sport and the outdoors, there was no prospect that his average football skills would land him a career in a major club. He had good maths results, was good at computers, learnt to cut code and the basics of the internet, enough to get a starter job in Brisbane or Sydney on a help desk. 'Only I don't want to move to the city, ' he'd sighed, dismissing the IT option as impracticable.

In fact it was Mrs Chalmers, the career advisor, who'd told him that the Northern Territory was recruiting for Community Police Officers. He reads in the prospectus

'Community Officers provide communication and liaison with local indigenous communities. They enforce the laws of the Northern Territory and protect and serve the public.

As an ACPO you are a role model and an advocate for cross-cultural awareness. You have an opportunity to make a difference and to improve community knowledge about policing services and law and order issues.'

Ben reads on. 'The ACPO training course incorporates 24 weeks paid, full-time training in Darwin. The course is demanding and involves physical training and learning legislation, operational safety tactics, driver training, information systems, communications and much more. After graduation, graduates will be posted to one of the Northern Territory's various police stations as a sworn ACPO. ACPOs will be subject to a probation period of 12 months from commencement.'

As the end of Year 12 approached, he'd put in an application. After the end of school assessment process, he'd achieved a Queensland Certificate of Education, and had in fact got a good score that would allow him a place in a number of educational institutions. The following March, he attended a comprehensive round of interviews with the NT Police, and was offered the job of probationary ACPO. He'd decided to accept.

Over the next few years, he'd risen to a number of challenges. He'd played the role of liaison officer between the people and the force. As an Aboriginal police officer, he learned people skills, and with plenty of encouragement, used his understanding of local indigenous cultures, languages, and customs, to build trust and rapport. He actually enjoyed acting as cultural liaison, helping to bridge communication gaps and promote understanding between the police force and the indigenous community, and found that his likeable nature and natural candour meant he was quite good at the role.

The job has plenty of challenges, where he feels he has to tread very softly indeed. He has walked a narrow path of getting on well with a largely white status quo, not noted for its flexibility in accommodating different points of views. He sees a genuine attempt being made by the establishment to deal with cultural sensitivities, language barriers, and complex social issues prevalent in the indigenous communities of the Northern Territory.

He learns to address matters related to family violence, substance abuse, land disputes, and community safety with tact and common sense. His cultural sensitivities acquired from family, give him enough understanding of traditional practices, kinship systems, and community dynamics enabling him to take a nuanced perspective, contributing to more effective outcomes, particularly when children and domestic violence are involved. He's seen enough, growing up, to take a practical peacemaker stance wherever the situation is not completely hopeless.

He finds himself in the position of being a role model and mentor for younger people. He tries to inspire the up and coming kids to take a positive attitude, and the wilder ones to step away from antisocial behaviours that would lead them to trouble. And he knows he is receiving valuable training and programs to develop his skills. The on-the-job experience as an indigenous officer is providing him with future options for career development, should he choose to move on in a few years time.

OK, he feels he still faces a lot of prejudice and preconceived notions based on race, but he considers he is in the fortunate position where his role is interesting enough, involving law enforcement duties, community engagement, and cultural advocacy. For the moment he feels that he can achieve some positive outcomes, while getting paid at a reasonable rate of remuneration. He feels lucky that this job is taking him in a direction he wants to go, which is at odds with many of his contemporaries, who are either stuck in low paid service jobs, or having to spend many years moulding themselves into jobs and career directions that seemed antiquated, outdated and even absurd to Gen Z.

And his post is at Jabiru, a town that he likes, not far from home in Katherine. All things considered, he reckons he's made a good career choice.

CHAPTER 18

Melanie looks up from listening to a podcast on her mobile phone, as she hears a vehicle pull into the car park. She is very pleased that it is Ben. She gets out of her car, and walks up to the driver's window. 'Thanks Ben. You sent me out to find a dead body. Not cool!'

'Sorry Melanie,' Ben sounds truly apologetic. 'I had no idea that would happen.'

'Come on. Let's get going.' Melanie makes a move. 'It's a pretty steep climb, and I do have to get to Jabiru.'

'OK, just show me what you've found, and I'll ask if you can go before they get here from Darwin.' Ben follows her towards the Gunlom Lookout Walk.

'I am not as fit as you. And I've already been up and back. So what was the report that you got?'

'It was called into the Jabiru station. I wasn't there, but when I got back from Darwin at the end of the day, the message was on my desk. So I called you and told you what was in the message. Which wasn't much. It was just that a car had been seen parked at the Falls, and whoever called it in, knew the Falls was closed. They said that they saw the closed signs, then drove back along the road to Boulder Creek. Because there was no-one in the car, they wondered if there might have been an accident. Then they drove back to Jabiru. They did their civic duty and called the station to report it the next day, because there was no mobile reception in the park. That was it.'

Twenty minutes later, they are standing by the body

on the rocks by the edge of the Falls.

'So where was he when you found him?' Ben walks around the body, looking at it from a few angles.

'Well, I didn't know he was dead. He was face down in the water, just here by the edge of the rocks. I thought he might be alive, so I waded in, and hauled him over onto his back. Then I could see he was clearly dead. It was horrible.' Melanie shudders.

Ben looks at the body of the man. 'His clothes look like he's a city boy. And his hands are soft. '

'I heard that Parks were sending up someone to talk to about Gunlom Falls. And I heard it was going to be Alex Petersen. I knew him when I was at uni in Canberra, but that was a long time ago. I can't be sure, his face is distorted, and it looks like he hit his head on the rocks. I think it might be Alex.' Melanie walks over to the edge of the Falls, looking out at the landscape. 'Ben, I need to get to the board meeting. Do you need me here?'

'You go Melanie. You've told me what you found. I'll wait here for the Crime Scene Examinations Unit. And I'll call it in that we don't need an ambulance. Catch up with you later.'

'Thanks Ben. I need a double latte and a doughnut. I think I'm still in shock.' Melanie heads back down the track, leaving Ben alone with the body.

Ben turns towards the view, and makes a call. 'This is Ben Mooney, Community Police Jabiru. Can I speak to the Senior Forensic Crime Scene Examiner, please?'

'Bell George? ' The voice is crisp and interrogatory when his call is put through.

'Ben Mooney, ACPO. I am at Gunlom Falls Infinity Pool. Just to confirm that it is a body here. It's a white male, mid twenties. He was found face down in the water, by Melanie Campbell from the Northern Land Council. She turned him face up, in case he wasn't dead. But he is. His face is swollen, maybe he's been in the water for a while, but I can see some marks on the side of his head. He may have bumped onto rocks in the pool. However it is possible he might have suffered a blow to the side of his head.'

'Identification?'

'I'm waiting for crime scene officers to get here. I haven't touched anything, haven't checked his pockets. But Melanie thinks it might be Alex Petersen from Parks Australia. She knew him when she went to uni in Canberra. I know that the Kakadu Board asked for discussions on what to do about the new stairs. I can call them to ask if they sent someone up here?'

'Do that, Ben. Then call me back.'

'Bob? It's Ben, Ben Mooney. Did they send someone up from Parks yet?'

'Yes. He's due to attend the Kakadu Board Meeting with Parks Darwin. Why?'

'Is his name Alex Petersen, by any chance?'

'Yes it is. Any reason why it shouldn't be?'

'I don't think he'll be turning up to that meeting.'

'Why not?'

'Because I think he's dead. I'm at Gunlom. Melanie found his body. Well she thinks it's his body. She's on her way to the board meeting by the way. She's had a bit of a shock.'

'She would. Are you sure he's dead? Because he was very much alive yesterday afternoon.'

'Thanks Bob. Look after Melanie. I'll let you know when I know anything.' Ben closes the call and hits redial to call Bell George, and tell her that it could well be Alex Petersen.

It is another few hours before the crime scene team arrives. Ben tells what he knows, then they set to work assessing the scene, looking around to determine the area to search for evidence. They set about the tasks of detailed documentation. Photographs are taken from multiple angles to capture the overall scene, with officers walking around the rock pools looking for specific marks on the rocks indicating potential points of entry into the water.

They carefully look for physical evidence in the pool where the body was found. DNA samples and fingerprints of the victim are taken. No blood is found on the rocks, but rock samples from potential entry points to the water are taken. The contents of the man's pockets are extracted. All the gathered evidence is packaged and labelled, sealed and documented to maintain a clear chain of custody at the scene.

Ben provides Melanie's contact details, explaining that she has continued on to the Kakadu Board Meeting in Jabiru. He provides as much information as he can about the report of the car sighting by the tourist.

'Can you let Melanie know we'll be asking her to come in to be interviewed when she returns to Darwin. You can tell her we'll want to see her even if we rule out a suspicious death. And can you interview whoever took the message at Jabiru Station to provide a report? Do you have a description of the vehicle? We'll need to find out if anyone else saw it. We don't know yet whether his head wound was caused by being pulled by the current down to the lower pool, or by a blow by a party or parties unknown. You could ask some questions in the community if anyone knows anything about a white vehicle.' Rachel, the senior Scene of Crimes officer on site, provides some advice to the young indigenous constable.

'OK. The vehicle description was just a white car, they didn't know the model or make, they called it in after they left Gunlom, because they thought somebody might have met with an accident. By the way,' Ben hesitates, then proceeds. 'Bob Wilson from Jabiru said he saw Alex Petersen yesterday. He's on the Kakadu Board, and he does the liaison with Parks.'

'If we decide it's definitely a homicide, we will take it from there.'

'OK. I'll head to Jabiru.' Ben sets off back down the walking track at a rapid pace.

CHAPTER 19

Melanie is back in her vehicle, starting the two hour drive to Jabiru, hopefully not too late to make the Kakadu Board meeting. She pulls out of the Gunlom Falls carpark, after what seems like a long time, but in reality only two hours since she first arrived. A couple of contentious matters are on today's meeting agenda, Gunlom Falls, and the clean up of the Ranger uranium mine. Deciding that she needs to focus, to take her mind off the alarming experience of finding a dead body, she finds the podcast on her phone from NITV Radio, the national indigenous station. 'Hopefully this will help me focus and forget,' she thinks as she clicks play, picking up from where she left off on the way from Katherine earlier in the day.

'Powerful words, sister'. The interviewer continues. 'I am speaking with Mirarr elder, Annie Lenton. Annie, what is your view? '

'We are first people. Some of our kids grow up to become black, but then they become white too. We gotta look both ways so first Australians and white people can live together. So we can all be together.'

'Do you think there is any hope of reaching an agreement with Parks Australia?'

'Yes. If they'll start listening to us. We've listened to them for a long time. But we've had a gutful. There was that helicopter crash a couple of years back. Parks Australia, they cut so many corners with their safety inspections. And now the government wants to go back on their promise, you know, the funding for the upgrade to the visitor centre? And fixing the roads. Promises they are not keeping. The Government said they were going to put in $200 million, but so far they only paid for $5 million. And Parks Australia, they blame Covid. We've got to ask them to fast track that money so we can make the place safe again. Then the tourists will come again. You've got to have roads if you're going to get here.'

The sound bite fades to the interviewer, who continues 'The Aboriginal Land Rights Northern Territory Act was closed to new claims in 1997. The Kakadu grants of land were conditional upon the Aboriginal land owners leasing the land back as Kakadu National Park. In the beginning this was a sit down between Parks and the traditional owners to share understanding, responsibility and decision making. '

'Not any more, Parks are just not listening,' thinks Melanie as she drives past seasonal floodplains, dotted with lush vegetation and surrounding wetlands covered by various types of aquatic plants, reeds, and water lilies.

'Traditional owners are looking at closing the popular destinations over concerns about a lack of respect, and increasing damage from invasive species. These people live, work and manage the World Heritage site. They closed Gunlom Falls, blaming Parks Australia for a breach of lease agreement, and lack of respect for sacred sites'.

Melanie is thinking about the last meeting between the Northern Land Council with the Mirarr traditional owners to protect Gunlom Falls, and discuss the redevelopment of the mining town Jabiru. They had not been happy with either Parks Australia as co-managers, or with ERA, Rio Tinto subsidiary, and their approach to cleaning up the Ranger uranium mine, surrounded by World Heritage Kakadu National Park. She turns up the volume on the podcast, as she rolls along, windows down, wind in her hair.

'Traditional owners say that nothing has changed since their scathing report on Parks management. They claim that indigenous traditional owners are still being denied jobs as rangers, and that Parks Australia are deliberately hiring outsiders instead of local young people.The clash has led to talks with Canberra to try to resolve the ongoing conflict between traditional owners and management. These talks have gone nowhere, traditional owners say. '

'When approached for comment', the interviewer continues, 'the Northern Land Council's Melanie Campbell provided the following statement - 'The NLC has a real obligation to act, not only because of the resolutions last week by the Kakadu Board of Management, but also because we've been receiving serious complaints for years from traditional owners about the mismanagement of the Park. Successive government ministers have claimed that they want resolution with the Board to resolve these long overdue matters. So far, on the ground, this has resulted in two ten minute Zoom calls, and a couple of vague promises. Attempts by the NLC to resolve these issues have been protracted, frustrating and difficult."

The podcast continues. 'Tourist numbers are up. However some of the major attractions are inaccessible because of poor management, and the failure by Parks to address transport infrastructure. Visitors have not been able to visit the Twin Falls for years. The Gunlom Falls, with natural infinity pools cascading over cliffs, is closed. Parks Australia built a new walking track over the periphery of a sacred site, against the backdrop of protest from traditional custodians, who are deeply hurt at the desecration. While Parks broke the law by disregarding the joint owners, proceeded to build the unsightly walking track, then lost the court case prosecuting the breach, the damage is already done.

Parks Australia have imposed what the Mirarr see as bad management practices, cutting the number of local rangers from the budget. When a fire broke out, millions of dollars of property damage ensued, lives were endangered, and staff cuts meant that no park rangers were available to fight the fire. Park ranger stations and houses were burnt to the ground, and traditional fire management practices were prevented because rangers lacked local traditional knowledge of Kakadu. The joint management arrangement between Parks and the traditional owners has really broken down in recent years. '

Annie gets her view heard. 'Parks Australia just stopped listening to us.'

The podcast continues. 'Canberra has been making decisions without consultation. There has been a serious deterioration of the environment from its pristine World Heritage status in only a few short years. Indigenous decisions have been overridden with arrogance and disregard for the situation on the ground. The Kakadu Board passed a vote-of-no-confidence in senior management, who resigned as a consequence, leaving a dysfunctional situation with no funds to redress the damage to the environment and buildings.

In the past, traditional owners were originally given land rights to be jointly managed with the Australian government. However the caveat was that the deal would not be honoured unless the local people agreed to allow uranium mining. Now the mine is finished. And the deal to restore the land to the same level as the surrounding environment, Kakadu National Park comes into effect. However, ERA, the Rio Tinto subsidiary responsible for the Ranger Mine wants to bury all contaminated waste in the mine site, with all the risks of permanent contamination of the surrounding ecology, particularly to the ponds and the Northern Territory wetlands. The local country is teeming with birds such as the jabiru and the brolga, wallabies coming to drink at waterholes, and long-necked turtles and estuarine crocodiles inhabiting the deep creeks and billabongs.

The miners want to renege on their financial contribution to restoring the land around the town of Jabiru, leaving the Australian taxpayers to foot a substantial part of the bill, and the people who live around and downstream of the contaminated water to inherit the problem.'

Melanie feels a surge of anger, at the thought of a public servant from the Canberra bubble being sent to pontificate to local indigenous people on behalf of the largely white government of the Australian establishment.

The NITV podcast continues. 'This is typical of the mine closures. The consequences of reaping billions of dollars in profit, is a legacy of environmental impact on the pristine ecology that can last for thousands of years or longer. While Rio Tinto's blowing up of Juukan Gorge is still fresh in the memory, the company might want to be seen to be accountable. But as soon as the fuss dies down, realistically, history shows that multinational mining companies do not take real responsibility for the damage they've done. It has been three years since the gorge's destruction. The Puutu Kunti Kurrama and Pedikura people have not forgotten their devastating loss. They have decided their best option is to work with the company to achieve co-management of country where they have been given leases. The question is, will Rio Tinto and ERA really clean up Jabiru? The answer is that so far they have shirked their responsibility.'

The podcast comes to a close, as Melanie travels the last fifty kilometres to Jabiru. She loves this place. The expansive landscape, plants and wildlife, stunning paintings of spirit beings, and wetlands filled with native and migratory birds filling vast tracts of floodplains make her heart sing with joy. Now she has to face the reality of the twists and turns of dealing with shady government politics, people with no love for the land, the environment or the well-being of the families who have been in continuous occupation in Kakadu. The landscape is degrading through mismanagement, threatening to be overrun by invasive species. Local people have been so hurt by past government interference and mismanagement, that they want to close the park altogether. A management crisis.

The wet season is over, the waterways are teeming with life, including crocodiles, a protected species and a past indigenous food source. She takes in the beautiful forested gorges, wetlands and the largest remaining intact tropical savannah unfolding on either side of the road. As always, her spirits rise as she passes through this wildly beautiful country, providing diverse habitat to hundreds of unique plant and animal species, now threatened by changes in the environment.

At last she passes the 'Welcome to Jabiru' sign, a metal outline of the black-necked, large white stork, with black bands across the wings, thinking that the real bird's neck is actually a beautiful iridescent purple-green.

She parks the car outside the Jabiru Community Centre, where the board meeting is to be held, met at the door by Bob. He tells her that with the news spreading about the body at Gunlom, they've postponed the meeting till the next day. Melanie breathes a sigh of relief. Bush Telegraph at work again. She needs some time to process the finding of the body, and what she has just seen at Gunlom Falls. Bob gives her the details of the room he's reserved for her at the town motel, and invites her to the evening's entertainment. It's a football match attended by a mix of the remaining mine workers, dwindling by the week, and local people. In the cool of the evening she is greeted by friends and relations at this popular social occasion at the Jabiru football ground. After the match, Melanie is glad to roll into bed in her overnight accommodation, to get a good night's sleep before attending the meeting the next day. It promises to be significant.

Interrupting a dream where soft voices, deep in conversation, emerge out of a backdrop of tall grasses growing under graceful eucalypts is the sound of honeyeaters and red-collared lorikeets, red-winged parrots, red-tailed black and sulphur-crested cockatoos forming a soft background chorus ringing through the landscape. No, it is a more strident tone, an electronic repetition. It is her mobile phone.

'Melanie Campbell?' she manages to feign a level of consciousness she does not feel. 'Melanie, it's Fabian. Fabian Bernard from UNESCO in Geneva again. I hope I have not woken you. I think it is after 8 am in the Northern Territory?'

'Fabian, nice to hear from you. Thanks for calling. Yes it's fine.' Melanie sits up in bed, and adjusts her thought processes away from the pleasant warm memory of a dream of country to business matter-of-factness. 'What time is it there? '

'I am at a meeting in New York. It's 6.30 pm, and we've just finished. I thought I'd call you to get your position on the Ranger Mine clean up, and how it affects the Park.

'Nice of you. Yes, the board is preparing a position to put to the Federal Government on what indigenous owners want from the cleanup. And the Northern Land Council has offered to help, with legal advice and any resources we can muster. The meeting was postponed till today'.

'Good luck with that, Melanie. We support your position. '

'Thanks. Successive governments have done the

damage. The underlying philosophy is do everything to accommodate the moneyed interests who support the party coffers, in the guise of being good for the economy. And nothing for sustainability. '

Fabian cuts in. 'And very little for the environment, which takes a back seat to outdated beliefs on the importance of money interests. Here in New York, we've been wondering if anything has changed since Australians went to the polls and elected the opposition party. Governments tend to avoid being held responsible for anything, particularly environmental damage, in my experience. They always seem to blame their predecessors.'

'Too true, Fabian. And the new government is blaming the old government. The housing market shambles has been exacerbated by past grants for homeowners to get upmarket new bathrooms and kitchens, and the Australian tax regime provides unprecedented tax benefits for owner landlords. The money in successive budgets for aged care is given directly to providers, many of them owned by overseas business interests, with little effective governance. And the government is downplaying the gravity of the global economic situation in a deliberate attempt to keep Australians feeling warm, fuzzy and safe so that out of gratitude they'll vote for them again. Sorry, I'll get off my soapbox.' Melanie is fully awake now, and appreciative of a chance to discuss the current situation with a friendly Swiss, who in his role as a UNESCO Environmental Capacity Building Program Manager, is empathetic to the problems of Kakadu.

'I think every country can feel the chill of a

looming deep recession, and a potential new pandemic outbreak.' There's a serious note in Fabian's voice.

'Yes. No doubt the governments will go to the next election with warm cuddly political advertising about their ability to save the public from impending disaster. No mention of super profits, tax cuts for the wealthy, greenwashing for climate change and the housing crisis. Who cares about the millions of Australians in the rental market? It must be their own fault for not buying up properties when they had the chance!'

Fabian laughs. 'Yes, jingoist messaging is springing up around the world. Australia is not the only country to experience a new nationalism.'

Melanie continues to air her frustrations. 'The Voice, an initiative supposed to address Closing the Gap for indigenous people, has increasingly become a slogan, and there is a lot of opposition. There is a token constitution change to recognise the first Australians, great, but it will still take us two or three years to shore up the legal position of land rights. It has been a couple of years now since Rio Tinto was given permission to blast Juukan Gorge under Section 18 of the Aboriginal Heritage Act.'

'Yes, as I understand it, Section 18 gives Australian governments the right to destroy your sacred sites on request from the mining companies?'

Melanie is fully awake now. 'Indeed. And they are doing virtually nothing about climate change, only green spin. When Europe, UK, China and the US were investing in renewables and new clean technologies, the major political parties here were still singing the praises of oil, coal and gas.'

Fabian provides a descant to her litany. 'And now the fossil fuel giants are singing the praises of Carbon Capture and Storage, while asking for new exploration licences. Commercial scale CCS is still a twinkle in the eye of optimistic scientists. They are doing something similar here in the US, as well as in the UK. Well I wish you the best for the meeting today. We are here to support you in your efforts to protect the Park. This includes any legal advice that you may need. I think you are right to get a strong position together to stop Parks from encroaching on any more sacred sites. And our understanding is that the Rio Tinto subsidiary ERA, responsible for Ranger, is trying to avoid their responsibility to clean up the uranium tailings. So please keep us posted, and let us know what we can do to help. We understand that the old clean up plan has been superseded?'

Melanie replies. 'Yes, the government has introduced legislation to guarantee and oversee the clean-up beyond 2026 and to ensure that ERA completes the rehabilitation. A step forward, because they were planning on doing a cheap, rushed job. But Rio Tinto is a big political donor to governments. ERA is asking for a new permit to prospect underneath the existing mine.'

'So the federal government is the new clean up overseer. Let's hope they do a better job than the past official cover ups, and the seemingly deliberate manoeuvres to make it too difficult for anyone to challenge poor clean up jobs. ' Fabian's no stranger to official obfuscation.

It is a relief for Melanie to be able to share her worries.

'Yes, that's what happened at Rum Jungle.'

Fabian is well versed in Australian mining history. 'Thanks for reminding me that there is a deeper uranium mining history in the Northern Territory than just Ranger.'

Melanie looks at the time, and brings the conversation to a close. 'I'd better get moving. Thanks for your call Fabian. I'll let you know any outcomes from the meeting. I think we may need some legal advice from UNESCO about the effects of the clean up on the park'.

'By the way, Melanie, I've got approval for a trip to Darwin. '

'Great! When are you coming?'

'I'll let you know as soon as I can. Talk soon. Maybe an online meeting next week? Bye for now.' As he closes the call, Fabian sighs out loud. He does not want to remind her that Kakadu's status as a UNESCO World Heritage site is in question.

Melanie did not think the time was right to share the finding of a body with Fabian. Not even when she is pretty sure the dead man is a Parks Australia public servant from Canberra.

She gets out of bed, showers and breakfasts in a hurry, before driving to the community centre where the board meeting gets underway. The first topic of discussion is the effects of the Jabiru clean up. The original plan was to bulldoze the town, but a proposal turned up to transform Jabiru into a tourism hub with five star accommodation, glamping and crocodile tours. But a realistic proposition? Melanie thinks that local people have every right to be sceptical of these grand visions, from bitter experience.

The Kakadu Board ponders the situation. How will politicians in Canberra deal with the clean up? With typical laconic humour, it is suggested that this may not be the most practical solution while the cane toads continue to advance, the fire threats escalate, and the weeds start to overrun the country and the ponds, threatening the wildlife. The board comes up with a resolution that Kakadu, home to people and their identity for millennia, must be protected from a poor clean up, and insensitive tourism development at Jabiru.

'You right bub? Ready to go? ' She's been invited to take the floor to discuss the second item of business, Gunlom Falls. Parks Australia has sent a representative from Darwin. Bob, Jawoyn traditional owner and board member, addresses Melanie directly. 'What does your mob have to tell us then?'

'Well, they think you have to go to the High Court. To decide whether sacred site protections apply. Because of the joint ownership, and because Traditional Owners were not consulted over the stairs at Gunlom Falls. '

'They built it right up to the men's business site. Women and children are not supposed to see that under our traditional law. We told them not to.' Bob's response is understandable.

'Well, there are constitutional issues. The Northern Territory supreme court has ruled that the director of Parks Australia is exempt from prosecution.' It's the Parks Australia board member from Darwin who provides this advice.

Bob interjects, 'We raised the case. We took your staff to the area and told them not to build there, and we also told them where they could build the path to the top. They just went ahead and spent millions of dollars of taxpayers money, and constructed the path right next to that significant area of the sacred site. '

Another traditional owner adds, 'The local tour operators are hoping for a boom in domestic tourism. But the infrastructure is falling into disrepair. Where are the government funds being spent?'

'So traditional owners want to take this to the High Court in view of the existing sacred site laws?' It's Melanie who asks this question.

'Yes. If the white man's law won't protect the land, we'll have to close parts of Kakadu. If the sacred sites cannot be protected, what else can we do?'

'The Northern Land Council will do whatever we can to help prepare the legal case. There are a lot of communities who are interested in this, because it will set a legal precedent.'

The vote is taken, and the decision is to appeal to the High Court. The meeting continues to discuss the items of business on the agenda, concerned with day to day running of the park. With the close of the meeting, board members pack up notes and laptops, and make for the door.

Bob catches up with Melanie on the way out. 'Are you OK, sister? You'd better get going. It's a long drive back to Darwin. Thanks for coming. Glad for your help'. He accompanies her to her car. 'Parks Australia might be saying they want to work with us now, but we all know that you can't trust them. They have history. They say one thing and mean another.'

CHAPTER 20

After a dinner of roasted vegetable and goat feta tart served with wild rocket and home-made pineapple chutney, followed by apple crumble and custard, the group is seated around the portable table and chairs, discussing the night's storytelling and stargazing.

Ron Jagamurra explains that he also studied European astronomy, before taking up a post in Melbourne to lecture on indigenous traditions. He tells the attentive group about the night's program.

'First I am going to recap on the information provided by the indigenous Warlpiri and Anangu guest speakers of the previous nights. Then I am going to explain what is to be seen through the telescope this evening. And following that, Mara has agreed to give a short presentation on the history of European astronomy.

As you know by now, the indigenous peoples of Australia have rich astronomical traditions that are deeply intertwined with their cultural and spiritual beliefs. While these traditions may vary among different indigenous groups, they all focus on using the night sky to convey stories, navigate, and mark significant events.

The Warlpiri people have a complex understanding of the night sky, with constellations forming an essential part of their storytelling and cosmology. Their constellations are often connected to Dreamtime stories, which are narratives that explain the creation of the world and its features. For instance, The Pleiades star cluster the 'Seven Sisters' holds cultural significance. It's associated with the story of seven sisters who fled from a lustful man, eventually transforming into stars to escape him. The Orion constellation is linked to a story of a hunter who pursues the Seven Sisters across the sky, and the Southern Cross and the Pointers, Alpha and Beta Centauri, are often used for navigation and orientation in the night sky.

The Anangu people have their own distinct knowledge of the stars and planets, often conveyed through oral tradition and rock art. They have a deep connection with the land, the sky, and their ancestral stories. The Milky Way is often seen as an emu. The dark dust lanes within the Milky Way are seen as the shape of an emu in the Anangu tradition. The Pleiades is also important to the Anangu people and holds similar cultural significance to other clan groups. Interesting that the Seven Sisters are the same in many cultures all over the world. Venus, the first celestial body to become visible in the sky at evening and is the last one to disappear from the sky at sunrise. Venus is revered as the Morning Star and the Evening Star. The same as in western astronomy, and we still don't know if the ancients knew it was the same planet.

These traditions have been passed down through generations and continue to be an integral part of the identity and worldview of communities. Details vary between different groups, and knowledge is shared orally. These stories you've heard on this course are just a sample of the constellations and celestial features and the deep connection that indigenous Australians have with the cosmos. Tonight's stargazing is to look up at the Milky Way's bright central luminous band of stars, stretching across the sky. This area is rich with densely packed stars and interstellar dust, creating a hazy and intricate tapestry of light.

Let's take it in turns to look through the telescopes at the various star clusters within the Milky Way. One notable example is the Omega Centauri globular cluster, which is one of the largest in the sky. It appears as a dense ball of stars. You heard from our guest speakers that to the First Australians, the dark regions are just as important as, if not more important than the Milky Way itself. In western astronomy, they are known as dark rifts or cosmic lanes. These areas are composed of interstellar dust which obscures the light from stars behind them. This creates a dramatic contrast against the brighter areas. Given the dark skies of a remote location, when it's your turn with the telescopes, you might see passing satellites as moving points of light. If you catch a glimpse of a meteor streaking across the sky, don't forget to tell the others. Enjoy.'

By now the group is a cohesive whole, and they do enjoy taking turns to observe, pointing out sky features that they have learned to recognise. At the end of the session, by invitation, Mara stands in front of the table, where they are all now seated with cups of tea, and large slices of cake.

She decides to explain a little of how the western system came about. 'The Southern hemisphere enjoys views of the Milky Way Galaxy which are geographically not possible in much of the Northern Hemisphere. The simple explanation is that we view space from Earth's perspective. The plane of the Sun's apparent annual motion is known as the plane of the Ecliptic. But the Sun does not revolve around Earth. So our view of the sky is how we see the sun, the planets, the moon and the zodiac stars that can be sighted from Earth throughout the year. We used to think that the sky revolved around us. Now we know better. So what we see has to be decoded into the movement of the solar system, and the movement of galaxies of stars.

Europeans found the zodiacal constellations particularly useful for determining the direction of east and west, and predicting seasonal changes. Ancient mariners observed the rising and setting of specific zodiac constellations to determine the time of year. This helped them plan voyages and know what weather to expect. So the stars told them the approximate east-west directions at night. For example, if they noticed that a particular zodiac constellation was rising in the east, they could use it as a reference point to determine their heading at sea. Mariners could identify specific constellations that were visible during specific times of the year and use them as fixed reference points in the sky to aid in their navigation. The angle above the horizon gives seafarers a rough estimate of their latitude in both the northern and southern hemisphere. That's how the Vikings made it to Iceland, Greenland and Newfoundland in Canada. Plus a bit of luck with the wind.

European navigation by the stars is really not very different from indigenous Australian practices. The real difference is in the cultural and mythological significance. And in fact, just like here in Australia, different cultures had their own stories and beliefs about the constellations. Mariners often used stories to remember the positions of the constellations. And while we know they realised that the night sky slowly moves, year by year, we still don't know whether they realised that the Earth was spinning around the sun with a slow precession around it's north-south axis. We know now that the Earth spins around on its tilted axis, returning to its original position more or less every 26,000 years, giving us a changing view of our place in the cosmos. They would have noticed that the seasons slightly changed alignment with the stars every year. And they noticed that the pole stars, the stars that are above the geographic north and south poles, are different over time. Did our ancestors work out that we are moving in some kind of pattern? I don't know.

When the constellations were named a few thousand years ago, Aries used to be aligned with the spring equinox. Because of Earth's precession, the spring equinox now occurs when the Sun appears aligned to Pisces. Ancient people noticed these changes, but how they explained them, we don't really know. By learning more about both indigenous and western astronomy, I think we can get some clues about how smart the ancestors really were.

Speaking of the equinoxes, the days and nights of equal length only occur twice a year. This is because the Earth's axis is tilted in respect to the plane of the Earth's orbit around the Sun. What caused Earth to tilt? Long, long ago, when Earth was young, it is thought that something big hit the planet and knocked it off-balance.

The constellations we see at night change from month to month due to the Earth's orbit around the Sun. And I'm pleased to tell you that the dark, clear skies of the Northern Territory provide a first class opportunity to explore the wonders of our galaxy through a telescope. Viewing in a remote location in the early hours of the morning is a truly breathtaking experience, due to the relatively low light pollution and the position here in the southern hemisphere. So I am going to grab this opportunity and stay up for a few hours of unique night sky viewing, if anyone cares to join me.'

Ron takes the floor again, supporting Mara's enthusiasm for the early hours view of the Milky Way, nonetheless reminding everyone that they are on the road by eight the following morning.

'Look, a shooting star!' A couple of hours later, Grace is first to spot the blaze of illumination from the transiting meteor above Earth, backlit by the iridescent night sky. 'What causes the light, Mara?' Useful to be stargazing with an astronomer.

'The heat causes atmospheric gases to glow brightly while the meteor passes through. The bright light tail lasts till the meteor burns itself out. Every few thousand years a really big one hits the ground, and causes some damage.'

'Sixty million odd years ago, a huge asteroid collided with Earth, sparking a mass extinction that wiped out most dinosaurs and three-quarters of the planet's plant and animal species,' adds Grace.

Most of their fellow travellers have gone to bed, tired

out after a day viewing the wonders of rock art, walking through lush monsoon rainforest, and swimming at Maguk Falls with its tiered, crystal clear natural pools.

'This is once in a lifetime for me,' Mara had declared. ' Winter moonless night in the wee small hours of the morning when the Milky Way is directly overhead. This is the best time to see the sky sprinkled with millions of bright stars. And this has to be one of the best locations for stargazing on the planet, because there is no light or atmospheric pollution, and no clouds. We don't even need a torch, we are illuminated from above.'

They are happy to just sit in silence, gazing upwards, with the occasional conversational gambit. With Mara in charge, they've still got one of the portable telescopes. Grace and Jonas, and a Canadian tourist have been taking turns viewing stars of interest. Suddenly the peaceful silence is broken, pierced by the rude noise of the engine roar of a vehicle travelling at speed, erupting from the road beside the campground. They scramble to their feet, and make out a vehicle passing a mere twenty metres from where they have set up the telescope. As quickly as it arrives, it vanishes, and the sound fades into the distance.

'That guy was driving too fast. And the noise. Oh now that is pollution,' Jonas remarks. The spell of the night sky is broken, and Grace looks at her watch. 'It's 12.30 am, time for me to get some sleep. We've got an early start tomorrow. ' The others agree, and helping Mara to carefully dismantle the telescope and pack it in the main tent, they disperse towards their individual tents, with a chorus of good nights.

In the morning, It is with real regret that they pack

up, stowing the gear in the trailers behind the two large four wheel drive vehicles, ready for the drive back to Darwin, and the afternoon debrief at Mindil. 'I'll remember this for as long as I live,' Mara comments as she, Grace and Jonas climb into one of the vehicles, after a last look around. They set out on the bush road back to the Kakadu Highway, leading onto the main road, the highway north.

CHAPTER 21

Yindi rises early and goes about her daily rituals of meditation and reflection, welcoming the rising sun. She likes to order her thoughts before beginning her daily routine. She feels the vibrant energy of the warmth begin to stir the country and its inhabitants. Birds are singing, tweeting and flying around on their morning expeditions. Bush mammals are scrabbling through the long grass. The air is cool, and she draws her possum skin cloak around her shoulders. She knows that there are changes under way in the climate. The rain is decreasing, and the lakes are beginning to be drier at the margins. She stirs the fire and takes a portion of the remains of yesterday's dinner after a successful extended family hunt. The roast kangaroo mixed with yams, flavoured with bush herbs tastes just as good as it did the night before. She thinks about last night's stories told round the fire, about the earlier times when there were plenty of kangaroos and emus, more rain, and rainforest that extended deep into the savannah grasslands.

She and her clan group live around and along the coastline. Walking out from the shelters around the campfire, where the others are still sleeping, she moves gracefully and silently along a well trodden bush path, surveying the surrounding country. By habit she observes the subtle signals that tell her where to go to collect the bush fruits, herbs and spices, and root vegetables that provide the accompaniments for the proceeds of the hunt. She makes a mental note of goanna tracks heading into the bushes, and decides to start the day collecting water.

She picks up the water carrier that she has made from the bark of a carefully selected tree, pliable and water-resistant. She made it herself, removing the outer layer of bark carefully, then folding the edges and sealing them with resin to create a watertight container. And now the water carrier hangs over her shoulder with a kangaroo hide strap. She moves towards the water's edge, the bank of the creek flowing out from a sandy edged billabong. She looks along the stream to the water flowing out to the sea, not so far away. 'Sing me your song, tell me about your journey. How are you travelling? Where have you come from? Do you always have enough speed to get all the way to the sea? Or do you sometimes stop and sleep for a few days, when the sun gets too hot? When you are waiting for the rain? Is there enough water for all your creatures, for all your trees and creepers along your banks?'

She smiles to herself as she remembers the stories, the experiences told with a mixture of drama and laughter over last evening's fire, stories that had lasted until everyone had gone to the gunyahs that provide protection during sleep, constructed using branches, bark, grass, and leaves. They shield against wind, rain, and cold temperatures, enabling a comfortable night's sleep. The clan has a collection of animal skins and furs from kangaroos, possums and emus used as blankets to provide warmth and insulation when the nights are cold. Walking along on this beautiful morning, Yindi remembers the stories of the stars told by her grandmother, about the sky beings who live in the blazingly bright and beautiful celestial heavens. She is worried however, about the way there is less rain, less water for the grasslands on the edge of the rainforest. She knows there is something different in her young life that she has to think about. The elders are telling the clan under the stars, by the fire, that the climate is changing. It is getting warmer.

And the rainforest seems to be retreating. She decides to think what to do if there is not enough water for everyone this season, including all the animals. She knows that when she has gone to the creek, there is a little less water than the same last time, same last season. She knows she has to do something, because her family trusts her judgement. She has been watching seasonal changes since she was born, as have all of her clan.

CHAPTER 22

Where there is an unexplained death, the NT Police go into action. The first responders secured the crime scene to prevent contamination and preserve evidence. They limited access to authorised personnel only. Not that there was anyone around in this remote part of Kakadu National Park, the end of the road, with the Falls closed to the public. The scene-of-the-crime officers completed the initial examination. They prevented contamination and tampering, preserved and documented evidence, and are packing up to return to base, where various authorities are notified, including the local coroner's office, and forensic experts.

They are treating the death as suspicious, undertaking the actions to investigate the circumstances surrounding the death, while the post-mortem examination is completed. For one thing, a positive identification has to be made. Following up on the suggestion that this may be the body of Alex Petersen, further investigation reveals that he has not turned up to today's scheduled meeting at the Darwin office of Parks Australia. Communication with Canberra reveals his accommodation, where he has not been seen for 24 hours.

Sandra Murphy, his senior officer, reveals to the NT Police that Alex comes from Darwin, and that in all probability, his parents still live there. This proves to be the case, and contact with Mr and Mrs Petersen provides the information that Alex was expected, and failed to turn up for dinner the previous day. With enough indications to show that there is a strong possibility that the body is indeed Alex, his parents are invited to do the identification. The body has already been taken to the Forensic Unit where the post-mortem and a forensic autopsy is to be conducted. At the Department of Health facility, Mr Petersen is escorted to see the body, and he identifies his son. Clearly upset, he emerges into the waiting area, where the news is shared with his wife, who unsurprisingly becomes distraught. The accompanying police officer does his best to comfort them.

The investigation wheels start turning. Next on the agenda for the officers assigned to a case, is to examine the evidence. The investigators, now that the body has been identified, start to gather information about the young man. They set about collecting identity documents, medical history, recent activities, personal mobile phone records, and start the process of tracking his whereabouts since his arrival at Darwin airport the previous Friday afternoon. They find the contacts he might have been expected to make since his arrival. The team, led by Detective Senior Sergeant Isabelle Felham, starts building the timeline, and scheduling interviews with Alex's family and colleagues at Parks Australia, both in Canberra and in Darwin, to gain an understanding of the events leading up to his death.

Isabelle is a senior-ranking detective who has substantial experience in criminal investigations. She has also been involved in major investigations into the drug trade, and quite a few murder cases. She is responsible for overseeing the work of other detectives, and coordinating efforts within the investigation team. Because of Alex's role in Parks Australia, the case has a reasonably high profile, so it's been determined that she reports progress directly to the Detective Superintendent.

She decides to start with a trip to Gunlom Falls. Booking out an NT Police standard issue Kia Stinger all wheel drive, she is accompanied by Rachel, one of the scene-of-crime examination team. Heading south out of Darwin, along Highway One, they reach the turnoff at Pine Creek to the Kakadu Highway leading to Jabiru. It's an all weather road in fairly good condition, so they make good progress. After reaching the Gimbat turnoff, they face a 35 kilometre side trip over lumpy and corrugated dirt roads to Kambolgie, Boulder Creek and finally Gunlom Falls at the end of the road. 'I haven't been here since I was at school,' remarks Isabelle, as they pull into the carpark. 'So take me through what you've discovered. I suppose there is nothing in the vehicle tracks that lets you distinguish one vehicle from all the rest. '

'No. Let's walk up the hiking track to the pools. Though we didn't find a lot there, either.' They set off up the bush track, and after some puffing and panting on the part of Isabelle, they reach the series of pools that spill down to the cascade that falls over the escarpment to the bushland below. 'Not as young as I used to be,' Isabelle pauses to take in the magnificent view across stone country and unfolding wetlands, as she makes an effort to slow her breathing. She turns to Rachel 'Now what did the team figure out?'

Rachel leads the way to the lowest pool, where water

spills over a shallow rock ledge to drop to the creek below, the largest rock water pool. As it's winter, the water has slowed from its wet season torrent to a gentle stream flow. 'This is where the body was found. We think he went into the water where the upper pool meets the lower pool, probably somewhere around there,' Rachel points to a site twenty metres away 'and then he was washed down, getting caught at the rocks directly above the falls.' She indicates the infinity pool, where water disappears over the edge into the vast Jawoyn country of southern Kakadu. 'Until the post mortem comes back, we don't know if he drowned, or if he hit his head, and the wound killed him before he went into the water. Or possibly his head bumped into a rock after he fell in. We haven't found any visible evidence of what could have caused the wound to the head. If he fell and hit his head while standing on the side of the pool, and then rolled into the water, there should be traces of blood somewhere. It has not been raining. There is nothing. And because the Falls are closed, the evidence cannot have been accidentally obliterated. '

'So what is your conclusion then?'

'If he hit his head on a rock, that rock would have to have been washed clean of any traces. We've covered the whole area with a fine tooth comb for signs of blood anywhere. The possibility is that he fell on a rock right at the water's edge with enough force to knock him out. That means he would have entered the water in the lower pool. And it is possible that he hit his head in very shallow water, meaning that all traces of blood and tissue have been washed away. We have not discounted the possibility that someone hit him with a rock or blunt instrument. In which case they've probably disposed of it. So until we get the post mortem results.... ' her voice trails off.

Isabelle moves back from the edge of the water. 'OK, I've seen enough, we'd better get moving to Jabiru. That's another two hours.' She turns and heads for the track, with Rachel following. 'Puzzling. So either hit his head on the rocks in shallow water, unless, as you say, he was hit over the head, and the weapon, rock or something else, was removed from the site. Which could have killed him, or he could have drowned.'

Having negotiated the dirt road back to the highway, and headed north, they finally pull into a parking spot in front of the Jabiru Police Station. Walking into the building, they approach the front desk. 'Detective Senior Sergeant Isabelle Felham here to see Constable Ben Mooney.' Ben is waiting for them, and shows them to a staff room with a table and a few chairs.

'We've just been to the Falls, we just need to check up on the details. Do you have anything to add to the information that's in your report?'

'I sent an update to the investigation team this morning.' Ben replies. 'Bob Wilson met Alex Petersen on Wednesday afternoon. Melanie Campbell thought the body might be him. Bob said this was at Alex's request, that he had called on Monday and made an appointment to meet him at the Falls at 4pm. '

'We only received the news this morning ourselves, but yes it is the body of Alex Petersen. And we've been on the road, so we didn't get your update. '

'Then you don't know that I tracked down the tourist who reported seeing a white vehicle there on Wednesday afternoon. He was staying at the Aurora Caravan Park. I called him on his mobile - he and his wife had already left, driving back to Queensland. He said he drove to Gunlom Falls, not knowing it was closed, on Wednesday afternoon. He and his wife did not see anyone there, just the white car standing empty, at around a quarter to four. Then when he saw the signs up, and the campground closed, he drove back to Boulder Creek Falls. They went for a short walk, and then because it was getting late, decided to drive on to Jabiru because he didn't want to drive along Gimbat Road in the dark. ' Ben is consulting his notebook, with the details he's jotted down. 'He said they got to Boulder around four. Walked as far as Plum Tree Creek, then turned back because it was getting too late to go all the way to the lookout. This was after they turned back from Gunlom. They set out for Jabiru around four thirty, and didn't pass another car till they got to the Kakadu Highway. He said his wife told him to report the car, because she was worried that someone might have been injured walking on the closed track. And there is no mobile phone reception.'

Can you arrange for Bob Wilson to talk to us? I'll give you my mobile phone number. We'll go to lunch, before we have to drive back to Darwin. If at all possible, I'd like to see him while we're here.'

'Ben, can you recommend somewhere to eat close by?' Rachel asks.

'Sure. The Kakadu Crocodile is OK. It's about two hundred metres to your left as you leave the building.

I'll give Bob a call, and tell him you're there. I told him to expect that he'd probably be interviewed.'

A half an hour later, Bob enters the restaurant, and joins Isabelle and Rachel as they are pushing back their plates. After the introductions, Isabelle remarks 'Will you join us for coffee? We can talk here. No need to go back to the station.'

'Just water for me.' Bob takes the proffered chair, and the waiter, having taken the order for two coffees and a mineral water, departs in the direction of the bar.

'So I understand that you met Alex Peterson on Wednesday afternoon.' It's Rachel who asks the first question.

'So it was his body at Gunlom?' Bob responds with a question.

'News travels fast.'

'Around here? Like lightning. Yes. Alex Petersen called me on Monday. Told me he'd been sent up from Canberra, to back up Parks in Darwin, and start new talks about how we can reopen Gunlom Falls. What he said was, Parks Australia wants to find out what we want.'

'And what did he say after that?'

'He said that he wanted to come here, to see Gunlom, so he could write a report on what has to be done. Asked if we could meet at the Falls, Wednesday afternoon at four.'

'So what did you say?'

'Yes. I said I'd meet him in the car park. I was there when he arrived. He turned up just after four.'

Rachel and Isabelle exchanged glances. 'And what happened then.'

'Well,' Bob continues. ' I showed him the stairs built by Parks, and pointed out how it is too close to our cultural site, and that we are not happy with Park's response so far.'

'And then what happened?'

'He went back to his car. When I left, he was sitting in the front seat. Looked like he was trying to make a call on his mobile phone. I could have told him there's no signal. Maybe at the very top of the Falls, but not in the Park.'

'What time was this?'

'Must have been about 4.30. I took him over to show him the stairs, and how they were in the wrong place. Then I showed him where the original walking track starts. He said that because it takes about twenty minutes walking up a steep track to the pools, Parks had built the stairs because they wanted to make them more accessible to older visitors. He asked me if we'd reopen the Falls when the court case is settled. I said no way, because we have to work out what's best to preserve the sacred site for the next generation first. Then we walked back to the vehicles. He said he'd get in touch after he put in his report to Canberra. He said Parks is interested in changing their attitudes, and being more receptive to what the Traditional Owners want to happen next. I said that we'd been ready to come to the table for months. Then I walked back to my car, and drove back home to Jabiru. He was sitting in his car, on his phone, when I drove off.'

'Did you see anyone else while you were there?'

'No.'

'Did you go up to the Falls?'

'No. We walked over to the track, and went up to the first bend. Took a look around, just to point out where the walking track is located, well away from any cultural sites, then came back down to the car park. '

'So when you say you got there at around four?'

'About ten to four. I was hoping to have enough time to have a good look around first, but I started out late.'

'And did you pass anyone on the way back?'

'No. There were no vehicles on the Gimbat Road. There might have been a couple of cars parked at Boulder, I'm not sure now whether I saw them on the way out to Gunlom, or on the way back. '

'What time did you get back to Jabiru? Did you stop anywhere on the way?'

'No, I went straight to Jabiru. Got home at about eight.'

'OK, Bob, thanks. Could you liaise with Ben Mooney, and go to Jabiru Police Station to make a statement in the next couple of days? We'll let you know when we're ready for you. At this stage it's just a formality. '

After his departure, Isabelle and Rachel discuss what they've just heard. 'We need the results of the post mortem pronto. We only have his word that they had an amicable conversation, and that he left before Alex. They could have both gone up to the pools, had an argument, and Bob could have hit him with a rock. '

'Or it could have happened how he said it did. He got there early, went for a walk up the track, and it could have been his car that the tourist saw, empty because Bob was out of sight up the track. Timing could work either way. ' Rachel responds with an alternative scenario.

Isabelle looks thoughtful. ' Do we know what colour vehicle he drives?'

Rachel replies immediately. 'Well judging by what he drove off in just now, it's a white Honda CRV circa 2001, the same colour that the tourist couple reported.'

'Motive?'

'Let's just say that there are a lot of very high feelings from the traditional owners about the lack of respect. Parks didn't even bother to consult them before building a culturally insensitive set of stairs for the tourists to take a short cut to the pools.'

'How come you are so knowledgeable, Rachel?' Isabelle is genuinely curious.

'I've been on the Indigenous Cultural Sensitivity course. You should try it. It was pretty good.'

'Maybe I will. We really need to see the forensic results now, to see if we can get a time of death. Come on, we've got a long drive ahead of us.' Isabelle walks over to the bar to pay the bill. 'That we have.' Rachel sighs, thinking it is she who has the long drive. They walk back to the police vehicle, Rachel gets into the driver's seat, while Isabelle opens the passenger side door.

The drive passes quickly and pleasantly enough,

particularly as they don't have to drive on the dusty, bumpy bush track from the Kakadu Highway to Gunlom Falls. Instead they can take the direct route back to Darwin on the well maintained Arnhem Highway. A couple of hours into the drive, almost to the outskirts of Darwin, Isabelle looks up from her tablet. ' The initial post mortem results are in.'

'So what does it say?' Rachel has her eyes on the road, with, however, enough spare attention to take in a high level overview of the findings.

'Well, OK.' Isabelle starts reading from the report. 'The immersion of a dead body in water can complicate the determination of the time of death. The water environment can have several effects on the body that can make the estimation of the time of death less precise, including temperature, decomposition, water movement and insect activity.' Isabelle pauses, scanning the document. 'Then there are a lot of technical explanations.' She scrolls through a couple of pages. ' Cutting to the chase. The indications are that death occurred somewhere between 5 pm and 10 pm Wednesday night. Cause of death is drowning. The gist of it is that there is a clear indication of a blow to the head, as there is residue of some kind of rock. Could have been before or after he went into the water.'

'Is that what it says in the report?'

'Yes. The blow didn't kill him, but it happened before he drowned. The force of the blow is such that it either had to be a blow to the head or a fall from a height onto a sharp rock. They just don't know because he'd been in the water for so long.' Isabelle is still swiping through the report. 'And he hadn't eaten since Wednesday lunchtime. OK, to sum it up, they say his death could be accidental, but they can't rule out foul play. It could go either way. He fell, hitting his head, went into the water and drowned. Or someone hit him with a rock and he fell into the water. Or he was pushed and he fell. But it still doesn't explain what he was doing alone up at the pools. If he was alone.'

So he'd been in the water about twelve hours before we arrived on the scene. So that rules out Melanie Campbell, not that I thought she was in the frame, but the timing still fits Bob Wilson if it's foul play.' Rachel offers ' And what is the final outcome?'

'The official result is 'Pending further investigation. Cause of death drowning, circumstances not fully determined'. According to Bob, Alex was in his car, on the phone when he left. ' Isabelle puts the tablet away, as they turn into Knuckey Street approaching the Darwin City Police Station.

'So in other words, he drowned, after receiving a blow to the head before or as he went into the water. And it is impossible to determine whether it was an accidental blow, or whether someone hit him over the head with a rock.' Rachel speaks slowly as she focuses on turning into the NT Police car park, pulling to a stop in one of the designated parking spaces. 'In other words, Alex's death may or may not be suspicious. Are they still doing analysis?'

'That about sums it up. And yes, they are analysing the wound to determine what kind of rock residue it is. They've sent samples of tissue to the lab. Meanwhile perhaps we can dig up some geological analysis of the kinds of rocks that are at Gunlom. That information must be somewhere in the system. ' Isabelle is getting out of the car and heading up the rear steps into the building. 'So clearly we have to interview Bob Wilson again. Can you line it up for next Monday? I have some matters to discuss with the Superintendent tomorrow. And ask Melanie Campbell to come in on Monday afternoon.'

'OK, Isabelle, I'll book the vehicle for Monday morning. I'll call Melanie at the Northern Land Council, and liaise with Jabiru to have Bob Wilson come in at 10 am. To give us time to get back from Jabiru, I'll ask Melanie to come in at two thirty. I'll get in touch with Parks Australia, find out where he was staying, and what kind of car he was driving. And I'll look into geological analysis.' Rachel also swipes her card, following Isabelle into the building, thinking that she has no wish to do another long drive the following morning either. And the lab results will take a couple of days to come back that will confirm or deny that Alex hit his head on a rock consistent with the geology of the pools.

'Ask the forensics unit to examine Alex's vehicle when you find it. And as Parks advised us he's driving a hire car, get in touch with the company and ask them about their tracking records. They should be able to tell you where he went. 'Isabelle throws this request over her shoulder as she disappears into her office.'

CHAPTER 23

Greg Crowther has had a chequered career since he left school, over a decade ago now. He is bright enough, born and raised in the city of Darwin, NT, with a career to date that has been characterised by ups, downs, and unexpected twists.

He enjoyed the three years he spent at Mindil University, enrolled in a course of Applied Science, studying a varied range of subjects, including biochemistry and eco tourism. Facing a choice between doing further studies down south, or getting a job, he drifts for a while. Nothing really appeals to him, however after a couple of months, tired of the economic constraints of unemployment, he lands a role with an adventure tourism company. He leads tour groups through Kakadu National Park. He is personable, and has an engaging manner. What he is able to tell the tourists about the ecosystem and Aboriginal culture is fairly limited, he makes up for it by telling amusing anecdotes about crocodiles and buffalo hunting.

While the job provides a steady income, Greg grapples with getting up early every morning, and after a couple of years, he tires of telling the same stories over and over again. He has kept up with his interest in music and the social events of Darwin. He manoeuvres a stint as a local radio host, where his personality and gift for storytelling keeps him in the gig for a couple of years. He enjoys interviewing people, and is able to talk about popular topics. He does not find it hard to echo the popular views of Territorians, because they are also his life views. They take him on as a permanent part time host for a feature show called 'What's On in Darwin'. He's messed around in boats for most of his life, so he is good at commentating on all things marine, sailing club events, boat races and community engagement. He provides a fast paced spruik of social occasions such as concerts featuring local bands, Darwin markets and beach clean-up events.

However, the radio industry is competitive, and over time he finds that there are younger people who are leapfrogging him into the more interesting program slots. He finds himself questioning the long-term sustainability of a career as a radio personality, although he determines to stick with it until something better comes along. It may not pay a lot, but he does not have to get up in the morning and don a suit, and he has a lot of time when he is not required to be researching and presenting 'What's On in Darwin'.

Greg has an entrepreneurial spirit - since his student days, he has dabbled in a little sideline, selling small quantities of hydroponic cannabis grown in the bush. He is supplied by a couple of old school mates who grew up on cattle stations, staying in town during the week while they attended Darwin High. He has been a nightlife habitue since high school days, and over the years he's developed a wide and varied circle of acquaintances in the clubs along Mitchell Street and in the Waterfront Precinct. He remembers one evening in particular, as he looks back on it. It was the evening he met Barry. He was sitting at the bar, having greeted and been greeted by a number of regulars, when a man whom he recognises as being in the circle of one of his old school friends, is sitting at the stool beside him, offering to buy him a beer. 'Call me Baz' he invites as he pays and lines up the beers. It is hot and humid in the Territory, and even with the air-conditioning on full blast, a cold beer mid evening is always a welcome offer, one that he accepted. Over a couple of months, it becomes a regular thing for his new friend to buy him a couple of beers.

Baz describes himself as being in the import/export business, and never accepts Greg's offer to return the compliment. And so it becomes a regular thing, that Greg gets a couple of free beers, while having a light hearted conversation about local topics of interest, and mutual acquaintances. Greg is never interested enough to find out exactly what Baz imports and exports. He is flattered that this clearly successful businessman has cultivated his acquaintance, and seems to find his company amusing and entertaining. He disappears for weeks at a time, then turns up again, saying he has been on an overseas business trip to Indonesia, Malaysia, Thailand or Cambodia, something that Greg finds impressive.

When after a few months Barry asks him if wants to earn some money helping him with a business deal, Greg is receptive.

'Look, it's like this mate. There is an opportunity to invest in an environmental sustainability venture in Bali. It's to turn local plastic waste into eco building materials. I really want to help the local community. But if I invest through my business, I am going to have to do a lot of paperwork, pay a lot of money to my lawyers and accountants, and set up a separate company. All of this takes time and money. These people want to get going quickly. They've lined up the equipment they need to buy online. If I go through my business accounts, I'll have to declare the transaction. And the Bali project would probably have to do a lot of paperwork and answer a lot of questions to the authorities, because they'll be dealing with suppliers in China. They just want to buy the plant and equipment, get it sent to them, set it up, and get the whole thing off the ground as quickly as possible. How would you like to help? There'd be a small commission and a trip to Bali in it.'

Greg is not particularly gullible, just a little naive. He could do with some extra money, and who doesn't want a trip to Bali? And he likes the thought of helping an environmental program to avoid some red tape, particularly dealing with plastic waste. Most sailors are disgusted with the amount of plastics that are so visible in the oceans and on the beaches. 'So what is the commission?'

'Do you think you could take the cash to Bali? Then they can split it up into smaller amounts, and put it into a few accounts, and use it to buy the gear online from China to set up the production line to turn the plastic waste into eco building blocks, without having authorities on their backs. Then they can get going in a matter of weeks, instead of months. You'd be paid, plus a free weekend in Bali. That's much cheaper for me than going through official channels.'

'Nothing illegal then?'

'No, not at all. They just ask how much cash you are carrying on the form at the airport, and you just put down under ten thousand dollars, and that's it. You get a weekend in Bali. They get the money in the bank to buy the equipment. And it's all done as a matter of trust.'

It all sounds attractive, so after a couple of weeks, Greg goes on a trip to Bali, handing over the package containing the money to an Indonesian man at the advance rendezvous set up by Baz, one of the busy outdoor beach cafes. He does try to ask the man about the plastic waste business, but he does not get very far, as there is clearly a language barrier, and with a lot of smiles and nods, they part and go their separate ways. The Bali trips turn out to be a regular event, and every three months or so, he is approached by Baz to do another trip. 'They need an extrusion machine. They are turning some of the waste plastic into packaging. They have to pay to get their process examined to be eligible for UN carbon credits. They need some replacement parts' are some of the very reasonable explanations that Baz proffers at their regular encounters over a beer.

It is around a year since they first met, and he is looking enthusiastic. 'Greg, I know you are a sailor, have you ever sailed a 45 footer?'

'What, a yacht?'

'I've seen a cruising yacht for sale on Bali. I've always wanted to sail. I do some offshore fishing in a mate's boat, but that's not the same as sailing the high seas. ' Barry is always a little flamboyant in his descriptive language. He pulls out a piece of paper from his pocket. 'One of my friends has sent me a photo and a spec for this boat. I like the look of it.' He passes the photo to Greg as he reads from the printed email.' 'Designed for comfortable and extended cruising, the Perihelion is a Bowman 45 yacht. It has a spacious interior with forward, master and guest cabins. U shaped galley, opposite the navigation station with a large chart table. Comfortable amenities for longer journeys. " Baz passes the description to Greg, who, after a quick scan asks ' so where is it moored?'

'Serangan Harbour. Could you sail it to Darwin, do you think?'

'If I had a crew, yes of course.'

'So how would you go about getting a crew?'

'Put an ad in an online sailing magazine. There are always people ready, willing and able to deliver a boat. Especially in Bali, if it means a free trip to Australia. No sweat.' Greg is now quite enthusiastic about the prospect. 'You'd have to allow about 10 days in case of bad weather. And you'd want to get going before the trade winds set in, otherwise you have to tack all the way.'

'So you're in?' Baz asks, reflectively looking directly at the young man sitting opposite him.

'What does it pay?'

'The usual, plus $200 a day allowance.'

'OK, but I have to take ten days off, so that's two grand for the trip, and let's say two grand for the time at sea.'

'OK, you drive a hard bargain. Oh there's one thing. One of my friends wants to send some sample eco biodegradable plastic sheets back, but he doesn't want them to go through customs, because there are lots of export regulations, paperwork, costs and local Indonesian officials who have to be paid. That makes it too hard, and too expensive to get sample merchandise to a potential customer in Sydney. You could make an overnight stop anchoring off Vernon Island, before you sail into Darwin Harbour. I'll come out to you in my mate's fishing boat to collect the package. Just makes it easier and cheaper all round.'

'OK. I can take ten days off with a week's notice. Are you planning on buying it straight away?'

'That I am,' Barry confirms. 'Can you put the notice in, and ask for the crew to be ready to go Friday week?'

'No problem, Baz, ' Greg assures him. There'll be plenty of girls and boys in Bali who'll jump at the chance.

CHAPTER 24

'I can't believe it's Friday afternoon. Where did the last two weeks go?' It's Mara who asks the rhetorical question. All the feedback forms have been filled in online. The short course is coming to a close, with thanks, exchange of contact details, and appreciation expressed to Ron Jagamarra for such a successful program. The field trip receives a five star rating, and the participants are starting to file out of the seminar room for the last time, with remarks such as 'excellent', 'best course I've ever been on' passed amongst the participants.

'What about another go at the Fiddler?' suggests Grace to both Mara and Jonas, all three, by now, being firm friends.

'I'm in.' Jonas is first to reply.

'Yes, me too.' Mara is only a couple of seconds behind with her assent.

'Same time, then? Seven?' There is a chorus of agreement. With a sense of satisfaction, and a tinge of regret, three people with very different backgrounds, from different parts of the world, walk out of the Mindil University for the last time, with an enduring bond of friendship forged through shared experiences.

'You've got Graham's contact number, Jonas?' asks Grace. 'Why don't we ask him to join us for dinner?'

'Good idea. I'll call him now.' Jonas pulls out his mobile, and makes a call. 'Graham? It's Jonas.'

'Hi Jonas, nice to hear from you.'

'I am here with Grace and Mara. It's the last day of our course. We are going to the Fiddler again tonight. Would you like to join us?'

'I would. One thing, I can't get there until around seven. I have an interview lined up with Melanie Campbell. I understand she was involved in your course?'

'Melanie. Yes. She addressed us a few times in our first week. Can you ask her if she'd like to join us too?'

'Jonas, I think that's a good idea. I'll ask her. I know she likes Fiddler's Green, because I met her there last Friday with her friends, after you left.'

'We were thinking of seven too. We can have a beer before we eat.'

'Sounds perfect. See you then, and I'll ask Melanie if she'd like to join us.'

'So Graham will join us, and he might bring Melanie?' Mara interprets Jonas' side of the conversation as they walk towards the bus stop back to town.

At the appointed hour, the Fiddler is as lively as it was the previous week, traditional Irish music pumping out, accompanied by a cheerful babble of conversation, something the Irish call the 'craic', an experience and feeling of having a great time.

Graham and Melanie are already sitting at the bar before the others arrive. 'So Fabian is coming out from New York, I think you said?'

'Yes, Parks Australia is immune from any type of

legal action. And Kakadu's World Heritage status is at stake. So the only recourse we've got to really protect Kakadu is the High Court. Gunlom Falls is likely to be a test case. The Kakadu Board has the numbers to support the traditional owners of Gunlom, the Jawoyn, even though Parks Australia is on the board. I've offered the traditional owners legal support on behalf of the Northern Land Council, but we certainly don't have enough funds to pay for the kind of representation they'll need to make their case. It is not easy to challenge when the Australian Constitution is involved. As you know it's our last chance to stand up for the preservation of Kakadu, and the management of tourist facilities in a way that is acceptable to indigenous people in the nation's highest court of appeal for civil matters. If we don't continue action, Parks Australia will just forget about doing anything for as long as it suits them.'

'So is Fabian offering any help?'

'Well UNESCO can't get involved in any appeal. Fabian is offering support for the legal team, which I think means they'll give us legal advice to build the case. Unofficially, of course.'

'That is generous. So they'll provide their international perspective, which can be passed on to your legal people. Because they are considering Kakadu National Park's World Heritage status. So when does he arrive?'

Melanie opens an email on her phone. ' In his last email he says he hasn't got a specific date yet. I get the impression that it could be in the next couple of weeks. He mentions that he has to visit Indonesia's newest site, in West Sumatra. Believe it or not, It's the oldest coal mine in South East Asia. The site has been turned into a museum. And the inference is that he could fly on to Darwin from when he's finished there.'

'Well you'll have to pull out all the stops to make some progress, because the tourist infrastructure is still deteriorating, and not being properly maintained. Particularly if Kakadu could be delisted. Call me cynical, but there is nothing we don't know about the views of the federal and NT governments. They pretend, but they don't really care about the natural environment, unless of course it involves natural resources exploitation. And Parks Australia is a federal government agency, answerable to the minister.'

'I can't call you cynical, Graham, because you're right. Look there's one of the astronomy course people. Mara, is it?'

Mara pulls up a stool, and having exchanged greetings, accepts Graham's offer of a beer.

'So what are your plans?' Melanie asks.'Are you staying around?'

'No, I only have a few more weeks in Australia before I'm due back at work in Edinburgh. So I'll head for Townsville, then down the coast to Brisbane, the Gold Coast and Sydney. That's all I'll have time for. '

Jonas and Grace are now in evidence, hovering at the bar. 'Shall I get a round? We can take it to our table.' Grace makes herself heard above the surrounding hubbub at the bar. 'I've booked us a table for 7.30. They're even busier than last Friday night.'

'I'll help you with the drinks.' Jonas offers.

After the move to the table, food orders are taken, drinks arrive, and the conversation turns to the astronomy course.

'The field trip was fabulous. Days spent looking at cultural sites thousands of years old. Incredible cave paintings. And nights stargazing in perfectly clear skies, with guidance from local astronomers. On the last night, when we were at the Kambolgie campground, Mara gave us a talk on the history of European astronomy. It seems like there is so much in common with Australian paleo astronomy. I learnt so much.' Grace's enthusiasm bubbles out.

'Maybe you should come visit us in Scotland to look at the night skies. In Aberdeenshire, there are over a hundred recumbent stone circles dating back to the Bronze Age. They have a huge stone lying on its side, the recumbent stone, with two flanking stones on either side. They mostly all face towards the south west, and they are built on low hills with clear views to the horizon.'

'So what was the purpose? I take it there was one.' Grace is clearly curious about stone circles.

'Good question. Well they were clearly for stargazing, solstices, equinoxes, sun and moon seasonal alignments, and local direction finding. Two of the stones often pointed across the landscape to the next stone circle. Bronze Age road signs, if you like. Cultural elements of course. Personally I get tired of archaeologists attributing everything in ancient monuments to spooky rituals, when primarily they were observatories. And likely gathering places and social venues for the local farming families. '

'I definitely want to visit Scotland, to explore the stone circles.' Jonas is enthusiastic. 'I loved Outlander,' he confesses.

'Well, I can't promise you'll fall through time. The stone circles form a wonderful horizon viewing frame through which to view the rising or setting of the major standstill moon that occurs roughly every eighteen years. The circles are designed so that the moon appears to stand still over the recumbent stone. '

'So what is a standstill moon?'

'Well. The moon's orbit of Earth wobbles, gradually causing it to rise at different points on the horizon over the years. ' Mara starts to explain.

'From what I've learned about astronomy,' she remarks, 'everything either tilts, inclines, wobbles, or spins.'

Mara laughs. 'That about sums it up. It does take a bit of getting used to. Anyway, if anyone wants to visit and get a fantastic view of the moon as it stands still, framed by huge standing stone pillars in a truly magical setting, then email me, and I'll take you to some of the best stone circles in North East Scotland.'

'I'd love to take you up on that, Mara.' Grace's enthusiasm is genuine.

'Me too. I'm interested.' Jonas takes a sip of his beer. 'Graham, Melanie?'

'Count me in.' Graham replies.

'If I happen to be on your side of the planet. When is the next Standstill Moon?' Melanie asks.

'2025. Not sure of the exact date, in January, if you're interested.'

'OK, Mara, I'm in. I am so counting on you to send out the invitation, the place, the time and the stone circle. I'll bring a bottle of best Highland malt whisky.'

'OK Grace. Consider it done.'

The dishes start to arrive, conversation is temporarily halted, and the table gets down to the business at hand. Once again the food is voted as excellent, without exception.

In the comfortable, relaxed feeling of repleteness after the meal, Graham breaks the silence, addressing Melanie. 'When I was at Jabiru today, I heard about you finding Alex Petersen at Gunlom Falls yesterday.'

'Yes it's true. It was pretty awful. I was there on Thursday morning. My cousin, Ben Mooney is a community constable at Jabiru. They'd had a report of a car parked there. Except that the Falls are closed. A man and his wife went there, found it closed, and drove back to Boulder Falls. The wife was worried someone might have had an accident, so she made her husband report it to Jabiru Station. Ben asked me to see if the car was still there on my way to Jabiru from Katherine. Well, Gunlom is one of my favourite places, and I was a bit early for a meeting, so I decided to walk up the track and look at the view. At first I thought it might be a croc. But it was a body. Apparently he'd died on Wednesday night.' Melanie looks troubled.

'I heard from Bob Wilson that the police think he was the last person to see him. He's worried that they might have him in their sights, if it turns out to be foul play. I was in Jabiru today to interview him about the Ranger clean up and the future of Jabiru. He told me that he had a talk with two policewomen from Darwin. They'd as good as said they're treating the death as suspicious' Graham addresses Melanie again. 'Have you heard anything more?'

'No. I'm expecting to be called in for an interview. I knew him when we were both students at the ANU. We both went to Darwin High. As a matter of fact, I saw him here last Friday night. He was with another man we went to school with, Greg Crowther.'

Grace questions Melanie. 'You were here last Friday? We were all having a drink. Jonas invited Graham to have a beer with us. I didn't notice you here, or I would have asked you to join us.'

'I arranged with Graham to meet him here after I had dinner with some friends. We had a lot to discuss about the situation at Gunlom Falls. It looks like the traditional owners have to take Parks Australia to the High Court, because it seems they have no liability or responsibility for actions they've taken, even though they are supposed to seek the consent of the traditional owners, the Jawoyn people. It seems like native title is overruled by the government whenever it suits them.'

'And Bob Wilson is a Jawoyn man?' asks Mara.

'Yes. He has been pretty vocal about his anger at Parks riding roughshod over his people's rights. So he's worried that he'll be a scapegoat for Alex's death. You see, Alex Petersen worked for Parks Australia in Canberra. Bob told me he said he was sent up to open discussions on the next steps for Gunlom Falls. Parks built it almost on top of a men's business site, without asking the Jawoyn.'

'We heard something about that on the course. Oh that's not good.' Jonas sounds indignant. 'When you say Bob was the last person to see this guy alive, where did he see him?'

Melanie answers his question. 'Alex asked Bob to meet him at Gunlom. Which he did. He said they walked around, looked at the stairs. Bob pointed out the direction of the men's business site, and they didn't go up to the site. He left him there, sitting in his car, on Wednesday afternoon, and returned to Jabiru. And the next morning, I found his body lying in the pool, face down.'

'We were in the neighbourhood the night he was killed then, ' Jonas responds reflectively. 'That's right, isn't it?' He turns to Grace and Mara.

'Absolutely. That was the last night of the field trip. You remember we stayed up late, looking at the planets and constellations that Mara showed us.' Grace confirms. 'And, that's when we heard a car that screeched by the camp at high speed, and disappeared back towards the highway.'

'You saw a car coming from the direction of the Falls on Wednesday night?' Melanie is suddenly interested.

'Yes. we could hear it coming. It was clearly travelling too fast. All the insects in the trees suddenly went silent. It was a shock to hear that kind of noise pollution in the middle of such a beautiful place.' Jonas tells what he heard.

'It completely broke the atmosphere of quiet

stargazing. In fact we stopped altogether, packed up and went to bed.' Mara adds.

'So did you happen to notice the time?' Graham asks.

'Oh yes, we had to be up early the next morning, it was nearly 12.30. I looked at my phone, and we decided to go to bed then.' It's Grace who answers.

Graham and Melanie look at each other. 'I think you'll have to tell the police. This might help Bob.'

'Do they know when he died?' Mara asks.

'No idea.' Melanie replies. 'He'd clearly been in the water for some time when I found him. If it turns out that he died late Wednesday night, or early Thursday morning, you might have seen the killer.'

'Did you actually see the vehicle?' asks Graham.

'Yes, the camp is right by the dirt road. There was no moon, but the stars were so bright you could make out the shape of the vehicle quite easily, even though it whizzed past us in a couple of seconds.'

'What colour was it?'

'Couldn't tell you that, but it looked like a dark colour.' Mara looks to Jonas and Grace for confirmation.

'As I remember, it looked like a four wheel drive, with a shape, like a Jeep.' Jonas offers.

'That's right, not a hatchback, and larger than a sedan.' Grace confirms.

'I think you'll have to go to the police station and tell them what you saw.' Graham looks serious.

'I've been asked to go in for an interview on Monday afternoon. Will you come with me, and tell them what you just said you saw? It could be really important for Bob. If Alex was killed late at night, that pretty well eliminates Bob.' Melanie asks earnestly.

'In terms of Bob, that's not exactly how the police will see it, however it's the best news we've heard yet to clear him from suspicion.' Graham adds.

'Yes, no problem.' says Grace. 'I can change my flight. And I can take an extra couple of days off. I just have to let work know. They won't mind when they hear how serious it is. They paid for me to go on the course.'

'I have a flexible bus ticket, I'll come,' says Jonas.

'I am here till Wednesday. Happy to go to the police with you.' Mara reassures Melanie.

'There's a cafe around the corner from my hotel. Passengers Cafe. I've been having breakfast there every day. Coffee's good. Why don't we all meet there for lunch on Monday.'

'Good choice.' Melanie replies to Grace's suggestion. 'And it is not far from the City Police Station. My interview is at two thirty. Can we meet at one, giving us time for lunch?'

'I'll come too,' Graham responds with firmness.

'I'll be there.' says Mara. 'Can you email me the details, Grace? Make sure I don't get lost?'

'Yes, me too.' replies Jonas. 'I am tired. I think I'll go back to the hostel, and get some sleep. It's been a long week.'

The others are in agreement, they call for the bill, and all make their way outside into the pleasant July tropical evening, saying their good nights.

CHAPTER 25

Alex's family wants answers. Despite the best efforts of liaison officers with NT Police that the jury is still out on what caused their son's death, Mr and Mrs Petersen are insistent that Alex was a victim of a terrible crime. While they are proud of their son's career in the national capital, and are generally supportive of the environmental activities of Parks Australia, they have a less than objective view of indigenous communities. When they hear he met a Jawoyn man at Gunlom Falls, they are sure he was responsible for what happened to Alex, either directly or indirectly.

Sandra Murphy from Parks Australia, closes the online meeting with Alex's parents with a sigh of relief. She just spent an hour of her life, reassuring Mr and Mrs Petersen that Alex was a valuable member of the Federal Government Agency's staff, and that he was not sent on a dangerous mission, alone in the Kakadu bush. In fact, she had to repeat herself many times. The grieving parents did not seem to take in what she was saying. That Alex was not in Darwin on a one man mission to face the anger of the traditional owners of Gunlom Falls. He was actually there to meet with his counterparts in Parks Darwin to discuss a number of matters to do with the Kakadu National Park. In no way was he representing Parks Australia to the aboriginal owners of the Falls, to the Kakadu Board, or to anyone else in the Northern Territory.

As Sandra explains in a call to Detective Senior Sergeant Isabelle Felham with the NT Police, Alex was on a routine trip to discuss Kakadu with Parks Darwin staff, who are the people who do the liaison with local indigenous communities. He was there to get an update on discussions with the Kakadu Board, and to attend the Board Meeting in the company of staff members from the Darwin office. And there were no special pressing requirements from the Minister, as he told his parents, that required Alex's personal attention. In fact, the Australian Parliament was not sitting, being on one of its lengthy breaks, so Alex was definitely not on a 'special mission' for the government, as he led them to believe.

Alex had been at work until Thursday lunchtime, and had not headed to the airport till mid afternoon Friday, under the Australian Public Service Commission flexible working arrangements which allowed him to work from home. He was expected back Friday morning, after talks and meetings at Parks Darwin office on Monday, Tuesday and Wednesday, and attendance at the Kakadu Board Meeting on Thursday. Sandra assures Isabelle that he was not asked to contact the Traditional Owners of Gunlom Falls, and certainly was not authorised to inspect the controversial set of stairs built close to a sacred site.

'Isabelle, it would be as much as my job's worth to send Alex to Gunlom Falls. A court decision has just been handed down in the Northern Territory, and we are advised that the Traditional Owners will be contesting the findings of no liability on the part of Parks Australia. Which means that the matter is still sub judice.'

'Understood, Sandra. So let me get this clear. Alex was

not back in the Northern Territory to speak to Jawoyn traditional owners of Gunlom Falls, on behalf of Parks. So what would his private mission have been?'

'That is it in a nutshell. No idea what he was doing meeting Bob Wilson there. Let me make it crystal clear. Alex went to Gunlom Falls with absolutely no authority to act in any way on behalf of Parks Australia in the dispute over the stairway. He was to attend the Kakadu Board Meeting with Parks Darwin, and report back to me on what was said and what happened. That was all.'

'OK, thanks Sandra. Alex's parents still think he was sent out to Gunlom for work. I am sorry you had to try to explain that to them. We'll be in contact with them as part of our investigation, so we'll try to break it to them gently, that he may have been acting privately.'

CHAPTER 26

The long-term investigation into drug running in the coastal north of Australia involves complex and coordinated efforts from multiple law enforcement agencies. The NT Police Drug and Organised Crime unit, known as DOCS, usually have a number of investigations going concurrently. When a lead, or a tip off comes to their attention, investigations begin with intelligence gathering. Officers try to identify what could be the potential drug trafficking networks, key individuals, routes, and methods involved. Research is conducted into various sources, including informants, and where locals are suspected of illegal activity, surveillance is conducted. Data analysts cross reference information to try to identify whether what they are hearing is worth following up.

These intelligence gathering efforts, even if there are no actionable insights, result in adding to the collection of information about suspected drug-related activities, individuals, and locations in the north.

Over time, they have built up a useful network of collaboration with informants, community members, and other law enforcement agencies. They are slowly gathering intelligence on the import of methamphetamines from Indonesia. Despite being assigned federal maritime and air assets for security operations, the Australian coastline is extensive, remote and wild. The northern border is known to be porous. Given the scale and complexity of drug trafficking, the Australian Federal Police is an important partner for DOCS. One of the current operations is a three way collaboration with Maritime Border Command of the Australian Border Force and the AFP.

They know that a surge of methamphetamines has been turning up in Sydney, since Covid lockdowns ended, and international borders opened. Analysis of the drugs reveals that they are similar to drugs that have previously arrived via well known routes from Thailand to Bali to Australia. Forensic chemistry techniques demonstrate that a lot of the supply is coming from just a few laboratories. Exactly how the drugs are leaving Bali is as yet unknown. The Indonesian Consulate in Darwin and the Australian Border Force have established a good cooperative relationship. And in a spirit of cooperation has led to the sharing of a tip from a Balinese police informant, about a new player who has been operating in the market recently, likely to be establishing a new route out of Bali. The intelligence is passed on to DOCS. Rumour has it that the new outfit is using various sea routes to get small shipments into Australia's north, before being transported overland to Sydney.

Initially, suspicion fell on routes from Indonesia through the Torres Strait, to Queensland's northern ports. Months of careful, painstaking research has so far not managed to identify in advance any shipment coming into the country. The Indonesian informant has now offered intelligence that Darwin is a destination, borne out by local meth supply increases. The transport is thought to be one of the burgeoning cruise ship tours that have increased in popularity, or private boats sailing from Indonesia.

It seems that the drugs business has its own informants, because so far, searches of boats as they dock in Australian ports have turned up nothing. Particularly disappointing is the failure of a concrete lead to a boat sailing from Bali to Darwin. The Perihelion was carefully monitored and tracked over the AIS marine navigation system as it made its way from Bali to Darwin. And carefully and methodically searched as it landed at its destination. Nothing.

Drugs and Organised Crime have been aware for some time, of a man who turned up in Darwin several years previously, purchasing an unimpressive apartment in Larrakeyah, not far from Mindil Beach. Despite living a quiet life, rumours began to emerge that he may have been a man who once plied the Sydney drug trade centred on Kings Cross, in the nineties.

The AFP, when showed photographs that had been taken surreptitiously in Darwin, compared it with photos on file for an elusive young man, suspected of being a courier for the notorious property developer, pubs and nightclubs owner, Abe Saffron, a man who virtually controlled the crime scene in Kings Cross, with various lines of business, including taking a percentage from drug dealers. At first they could not be sure if Perron is the same man who turned up in Darwin, clearly in his fifties, whose features are difficult to compare with his younger self, if indeed that is who it is. He disappeared from surveillance at a time when Saffron's influence was in decline, and the drug scene in Kings Cross was passing into the hands of an organised crime group of northern Italian origin from the inner western suburbs.

The NSW police knew that Barry Perron was probably a courier, travelling to and from Thailand, however he was either never exposed enough to come to serious attention, or more likely, as one of his schoolmates had joined the police force and ended up stationed in Darlinghurst, then notorious for corruption, he greased the wheels of his freedom and anonymity. The man living in Darwin is also called Barry Perron. Immigration records reveal that he left Australia in 2001, only returning in 2017. No earlier records were found, probably because Australia only started keeping permanent passenger records in 1999.

On his return to Australia, he flew straight to Darwin, buying an unprepossessing apartment close to the city and Mindil Beach with cash that he legally transferred in from a Swiss bank, according to Austrac, the agency responsible for preventing and detecting financial fraud. Their records show Perron declared the money as proceeds of overseas earnings on investments. The AFP noted in its profile dossier that it may have been as a precautionary response to Swiss banking secrecy ending in 2018. Also in his file is the fact that since 2017, Perron has made regular trips out of the country, particularly to Thailand, Cambodia and Vietnam. Again nothing suspicious to report.

Background checks revealed that he was born in 1969, grew up in pre-gentrification Surry Hills, went to school at Sydney Boys High School, after which he does not show up on government records until 2001. He has never shown up on Australian Tax Office records, so clearly, when in Australia he supported himself by independent means, which was possibly the drug trade.

With nothing further to go on, DOCS keep an eye on him, as a regular in the night club scene, but they do not see any activity that would have led to suspicion. The odd coincidence is that he appears to be friendly with Greg Crowther, the same young man who skippered the yacht Perihelion from Bali to Darwin, crewed by backpackers. The same boat, registered to an Indonesian man, about which Indonesian border forces had notified the AFP as being worthy of investigation owing to a tip off which led nowhere.

This triggers more activity on the part of DOCS Darwin, and the task lands on Jen's desk, an analyst of less than two years standing. She's a recent graduate, still keen to make a difference, so she sets about remotely accessing scans of reports from the 1990s about the drug trade in the Cross. She proves to be unusually hardworking and diligent in pursuit of answers as she tries to piece together Barry Perron's past.

The questions are fairly straightforward. Did Perron work directly for Saffron? Did he work for one of the Kings Cross drug dealers who gave Saffron a cut? Or did he work for the incoming drug dealers taking over the turf? Did he operate with members of Darlinghurst police, several of whom, it is acknowledged in policing circles, were known to be bent, turning a blind eye to the drug trade? Jen finds it illuminating, and after a couple of days reading between the lines of old reports, she suspects that the corruption went way beyond looking the other way.

Another question that springs to Jen's mind is that

of a couple of detectives who seem to have changed allegiance to the group from the inner west, were gradually taking over from Abe Saffron's network. A bit of detective work of her own reveals that the incomers were connected to a Northern Italian family, who having moved north from Sicily early in the twentieth century, emigrated to Australia in the seventies. After trawling through many typed reports about surveillance of the Saffron connected drug network, she finds a single mention of Perron as a possible bagman. This was soon after the new mob was starting to establish itself in Kings Cross. 'And in this case,' she reflects, 'mob really does mean family, a particular kind of family.' Jen infers that Perron may have, along with his police connections, changed allegiance, once the writing was on the wall. Not a comfortable position to be in. No drug faction would have tolerated disloyalty, so if he did change allegiance, it would only be a matter of time before he would be caught in crossfire, perhaps literally.

So any sensible young man with no ties, no kids, no mortgage and no other career options, would take the best option available to ensure a long and healthy life. Perron left the country. If it ever got out that Perron was muscling in on overseas suppliers, he would not last long in Sydney. He probably could only keep a low profile for a short period of time. In 2001 he left the country. So where did he go? Did he leave the country with cash from a drug deal? In which case he would have to completely disappear.

Jen is sure she has enough reasons to make a wider search, so she makes a case to her superior officer in

DOCS to reach out to Interpol, looking for confirmation of her suspicion that Perron spent his years abroad after he left Sydney, involved in the drug trade, judging by the $5 million he legally brought back to Australia. The approvals process and the request itself takes time, so she busies herself reaching out to colleagues in NT police for intelligence sharing about Perron. Which throws up his apparent friendship with Greg Crowther, and the tip off about the Perihelion, skippered by Crowther, and the search in port that found nothing.

CHAPTER 27

Isabelle and Rachel are at Jabiru Police Station, recording the interview with Bob Wilson. Formalities over, they pick up from the previous week's conversation.

'Thanks for coming in, Mr. Wilson. We just have a couple of questions for you. You say you left Gunlom Falls carpark at around 4.30 pm, and that the last time you saw Alex Petersen, he was sitting in his vehicle. '

'Yes that's right.' Bob answers promptly.

'What kind of car was he driving?'

'Late model four wheel drive, I think it was a Holden Calais, though I couldn't be sure. It was maroon.'

'So, Mr Wilson, you recall that you stated on Friday that you got back to Jabiru at around 7 pm. Is that correct?

'Yes'

'Enquiries made by Jabiru police show that you called in at the Kakadu Crocodile.'

Isabelle consults her notes. 'And, the proprietor, who was on the front counter on Friday night, remembers you coming in. He tells us that you said you'd just got back from a useless trip to Gunlom Falls to talk to Parks Australia'

'That's right.' Bob addresses the women opposite him.

Rachel pipes up. 'He said you came in just before

eight. To ensure his memory was correct, we got him to check the till receipts. You ordered a beer, and a Smoked Roo Salad which you ate at the bar.'

'That's right.' Bob wonders what's coming. He can sense a follow up.

'You said on Friday that you got back around 7 pm. So what did you do with the extra hour?' It's Isabelle who goes in for the kill.

'Yes, that's right. It took me two and a half hours to drive back from Gunlom. I went to the office, which is at the Community Centre, to fill in the travel diary. I have to record mileage, date, time spent on the road, and the reason for the trip. Then I went to get some dinner. When I go on a trip for Kakadu Park business, I get paid for petrol, and a meal allowance when it is over 100 kilometres. We get paid the same travel allowance as public servants. I have to fill in my travel details online. ' Bob replies.

'Can anyone confirm that?' Rachel shoots back.

'No. They'd all gone home. After hours.' Bob thinks that they are not done yet, so prepares for the next volley of fire.

'So it took you the best part of an hour to fill in a travel diary?' It's Isabelle this time.

'Yes. It always takes me a while to fill in online forms.' Bob feels he's parried their thrusts quite well. He is no stranger to dealing with NT Police questions.

Isabelle and Rachel exchange glances. 'You'll need to provide proof about when you filled in the travel diary .'

Bob looks at them blankly. 'How am I supposed to do

that?'

'Ask whoever does the Kakadu Board IT to print out confirmation of when you filled in the diary, and get it to us as soon as possible. All right, you can go now, Mr Wilson. We'll want to talk to you again. Oh, one more question. Did Alex Petersen seem drunk when you met him?'

'No.' Bob has had enough questions, and he rises and makes for the door.

'Interview ended at 10.35 am.' Isabelle presses stop on her mobile phone.

After Bob's departure, Isabelle and Rachel confer. 'Couldn't have taken more than twenty minutes to fill in a travel diary. I don't believe he's fully accounted for his movements.' It's Rachel who speaks first.

Isabelle has the grace to look thoughtful. 'Hmmm. I must say, it usually takes me ages to figure out my online travel diary. It usually takes me a while to work out what to put where, and then I have to write up the trip in official language. Rachel, we'll let it drop for the moment, until we get the analysis of Alex Petersen's head wound. If it turns out the type of rock is not found at the pool, then we'll do a forensic examination of Bob's car. Let's see if Bob manages to show some evidence of when he filled in his travel diary. '

'OK. In that case, let's get a coffee and a sandwich to take away, and get back on the road to Darwin. Don't forget we are interviewing Melanie Campbell this afternoon.' Rachel leads the way back to the Kia Stinger.

CHAPTER 28

Passengers cafe is a bright open space, with dark wooden furniture that lends a hint of old colonial decor, lit by large plate glass windows, allowing a wide view of the streets outside. Mara steps in and sniffs appreciatively at the rich aroma of freshly brewed coffee. A young woman with an easygoing manner stands behind the counter, and after a few friendly exchanges, she delivers a latte, and a sandwich. Looking around, she finds that her friends are already installed at one of the tables, and she wanders over to join them.

'Sorry I'm late, ' she explains as she seats herself in the spare chair. 'I decided to walk, and hadn't realised that it is such a long way from the house where I'm staying.

'No problem, good to see you. We decided not to wait, so glad you've got something to eat.' Grace's response is friendly and welcoming. 'We've started discussing the interview, and we've decided that we'll just turn up with Melanie. And explain that we have something important to relate about what we heard and saw on Wednesday night. Or rather early Thursday morning, actually.'

Graham clarifies that he and Melanie have determined that he'll go along to the interview too, to relate that he saw the two men at the bar of the Fiddler, whom Melanie identified as Alex Petersen and Greg Crowther. After he parted company from the chance encounter with Jonas and co, he waited for Melanie at the bar. 'I didn't know who they were, but I clearly recall seeing two men seated at the other end of the counter, while I waited for Melanie to turn up. They remained there for the whole time we were discussing the next Kakadu Board meeting.'

Melanie confirms 'They were definitely Alex Petersen and Greg Crowther. They were sitting there when I arrived to meet Graham. I went to school with them, so there was no mistaking who they were. ' She glances at her phone. 'We'd better get going, it's a ten minute walk to City Police. I'll be glad of your company.'

After explanations at the front desk, Rachel emerges from behind the security area to hear Melanie's explanation that her friends have further information about Alex Petersen. After disappearing to consult Isabelle, she returns, inviting the group to accompany her to an interview room, where Isabelle is already seated, albeit a little grumpy as they've not had time for lunch after the long drive back from Jabiru.

'So Melanie, your friends have some information about Alex Petersen from their astronomy field trip? ' She turns to address the group who are being ushered into the interview room. 'We are certainly interested to hear what you have to say. As you are volunteering to help us with our enquiries, I take it you have no objection to our recording what you say? Let's start with what happened at Kambolgie Campground.' Isabelle sits back, inviting the three field trip participants to tell what they know. They recount the late night incident, including what they recalled of the vehicle. 'You think it might have been a Jeep?' Rachel asks.

'No. It just reminded me of a late model Jeep. Not the early ones. You know, rounded off top for off road driving. Could have been a Suzuki or a Toyota, we just didn't have enough time to see the make and model. It was gone in a flash.' Grace responds.

'And you don't know what colour it was?' Isabelle continues.

'No. Just that it was a dark colour. The stars were so bright that we could see it wasn't a light colour. I don't know if it was dark blue, dark red, dark grey or black. It went by so fast.' Jonas adds. 'There was enough light to see it wasn't white or silver, that would have stood out.'

'We were taking it in turns to look through the telescope at Saturn in Aquarius when we heard the vehicle approaching. We all stopped and looked towards the road, which runs right by the part of the campground where we had the telescope set up. We watched as it roared by. It was travelling far too fast. Like a bat out of hell. but we didn't think much about it until we met up with Melanie and Graham at the Fiddler, and heard about the guy dying at Gunlom Falls, a few kilometres away.' Mara speaks for them all, as she recounts what they observed when the speeding vehicle intruded on the peace of the Kakadu night. 'It was a bit unnerving, so we decided to call it a night.'

Rachel nods, taking notes. 'So did you get a good look at the person in the vehicle? '

'Couldn't see them clearly. Just that it looked like there was only one person. That was the impression I got.' Mara frowns as she concentrates. 'Honestly, I couldn't swear to it though.' Grace and Jonas nod in agreement.

'Yes I thought it was one person too.' adds Grace.

'Thank you, Jonas, Mara and Grace. This is valuable information,' Isabelle says sincerely. 'Glad you came forward.' She looks to Graham, whom she knows of, by reputation, as a journalist who asks awkward questions. She decides to treat him with kid gloves. He provides his corroboration that he also saw two men at the bar, for the whole time he and Melanie were talking at the Fiddler. Melanie attests that she identified these same men as Alex Petersen and Greg Crowther, whom she recognised from school days at Darwin High, and they were still there when she left, at the same time as Graham.

Graham offers that they appeared to be in deep conversation. No he had not overheard what they were saying as they were at the other end of the bar, with their heads close together, and besides the bar was noisy. Melanie agrees that this was also her impression. Isabelle politely dismisses the four extra people from the interview room. 'Can you leave your contact details at the desk as you leave? Thank you. Your information may prove helpful.' Rachel accompanies them to the front desk, while Isabelle turns back to Melanie.

This time, it's a formal interview, and Melanie relates the finding of the body. As the interview concludes she feels a sense of gratitude that people from different parts of the world had taken the trouble to provide an account which must have a bearing on the death of the young man from Parks Australia, and would probably clear Bob of suspicion. As she leaves, she is pleased to find Graham and her new friends waiting for her in the reception area of the station.

'Come on, let's go have a cup of tea. We can exchange

contacts in case you need us again, if it'll help.' Grace's suggestion seems like an all round good idea, so they walk back along the city streets to Passengers.

'Give me your email address everyone. So I can invite you to Standstill Moon events at the stone circles of Aberdeenshire.' Mara opens her phone to her contacts, over tea and scones.

When she returns to the interview room, Rachel and Isabelle exchange notes and impressions. 'I think there is a strong possibility that the vehicle they saw is involved in Alex's death.' Rachel says what she thinks.

'Clearly our next move is to interview Greg Crowther. Particularly in view of the fact that Parks are telling us that Alex had no business to ask Bob Wilson for a meeting at Gunlom Falls. If indeed he did. I've already asked Sandra Murphy from Parks for a copy of Alex's emails and a record of calls made from Alex's work mobile. I still have my doubts about Bob Wilson, but he's looking less and less likely. I wonder why Alex wanted Bob to show him the stairs at Gunlom Falls. I think we still don't know everything about that.' Isabelle sighs deeply. I have to see the Super this afternoon, and I am going to have to tell him that we are pursuing lines of enquiry, but we now have to follow up with a new lead. Can you find out what you can about Greg Crowther, and let me know as soon as possible, so I'll have something concrete to take to my meeting this afternoon? I'd like to be able to say we are getting somewhere.'

'OK, I'll do what I can,' Rachel is glad that it is not her who has to account for progress up the food chain in what has turned out to be an uncomfortably high profile investigation. 'What about Melanie? She said she went to school with both Alex and Greg.'

'Yes but Graham Saunders confirms that he was with Melanie in his professional capacity to discuss the Kakadu Board Meeting. And they say they left together, so it probably happened just as Melanie said. She happened to see them at the bar. Which in fact is helpful, as we don't have any other leads to go on.I think we have to direct our enquiries in Greg's direction.'

'OK Isabelle. I'll find out as much as I can about Greg as quickly as I can. Let's see if I can find something in time for your meeting with the Super. ' Rachel is sincere in her intention to help.

It's only an hour later that Rachel knocks on the door of Isabelle's office.

'Think I've got something for you.' Rachel moves into the room unasked and takes a seat. 'I ran a check on NT Police systems, and when I text searched Greg Crowther, I got a hit in this DOCS report.'

'Drugs and Organised Crime?' Isabelle is looking interested now.

'Yes, it's in connection with a drug investigation that came from a tip off from Bali police. The original source was an informer. The tip was that a boat called Perihelion may be containing methamphetamines. Now Greg is known to be an associate of a man named Barry Perron, who came to the Territory in 2017. It turns out that he was known to police in Sydney in the nineties in connection with Abe Saffron, a notorious Sydney property developer and crime boss, who died in 2006 at the age of 87. In the early 2000s, there was a takeover of the Kings Cross drug business from Abe Saffron and associates, by a new group from the Western Suburbs, with links to Northern Italy. Around this time, Perron went overseas, and did not come back at all until he landed here in Darwin in 2017.

Jen from DOCS has sent a request to Interpol looking for information on what he was up to overseas, but she's heard nothing back as yet. And here's what is really interesting. It was Greg Crowther who skippered Perihelion from Bali to Darwin a few months ago, crewed by a bunch of backpackers who, as far as we can tell, responded to an ad for free passage to Australia. The investigation is ongoing. Jen, the analyst who's been assigned to trawling through old Sydney case notes that were scanned and put online recently, has promised to keep me posted when new information comes to light. Oh and one more thing, DOCS have had both Barry Perron and Greg Crowther under surveillance for the past month. They meet every couple of weeks in one of the nightclubs. '

'Great. I've got something to go to the Super with. Thanks Rachel, really appreciated. Now we just need to find out what business Alex and Greg were discussing at Fiddler's Green ten days ago. And who drove a four wheel drive past Kambolgie campground in such a hurry last Wednesday night.'

'I'll start by checking what vehicles Perron and Crowther drive.' Rachel's closing remark is made as she leaves Isabelle's office. 'And of course Alex's hire car.'

CHAPTER 29

Sandra Murphy has provided a detailed account of her dealings with the Northern Territory Police to the senior management team of Parks Australia. It doesn't make comfortable reading, no matter how she glosses it over, that Alex Petersen has made an unauthorised and possibly illegal trip to Gunlom Falls under false pretences.

She is expecting the meeting, and her expectations are realised when she is called to provide a full and frank account of what she knows, and what she has learnt about how Alex Petersen came to get in contact with traditional owner and Kakadu Board member Bob Wilson, and whether there was any indication that he intended to visit Gunlom Falls.

The meeting is as sticky as she expects, and it ends with a promise on her part to talk to all of Alex's colleagues to see if he confided his motives to anyone. A follow up meeting is scheduled for the next day to see if she can shed any new light, or gain a new understanding of the young man's reasons.

Clearly she has to pull out all the stops, so she calls an urgent meeting of everyone he worked with who had day to day contact with Alex. His work colleagues have a predictable series of responses, including 'Terrible news about Alex', 'What happened to Alex?', 'Was it an accident?', 'How did it happen', and 'OMG what was he doing at Gunlom Falls?', the latter from Anna who worked most closely with him on a daily basis.

In response to Sandra's question of anything unusual that Alex had talked about regarding his visit to Darwin, it is Anna who has a response. 'He said he was going to try to find out about something he heard about from his friend at the Indonesian embassy, Safira I think her name is.'

'And what did he hear from this person, Safira?' Sandra's suddenly very formal and serious in tone. 'And exactly who is Safira?'

Anna sees no reason in keeping Alex's confidences any longer. 'He met her when he went to the Malaysian Embassy event with you, Sandra. She does Media Liaison. She and Alex go, went surfing together at weekends. With someone from the Malaysian embassy. Alex was learning to surf at Broulee whenever he could. He invited them to go with him, and they learned to surf too.'

'And how often did they go surfing?' Sandra's voice is calm.

'Every couple of weeks, I think. He said that they both liked surfing, and that they liked each other. It was a chance to get together and get to know each other better. '

'You said he found out about something from Safira? What exactly did he find out?.' Sandra is clearly making an effort now to keep the edge out of her voice.

'Just that her aunt owns a hotel in Bali, and a chain throughout Indo, and that she'd been approached to build a tourist facility at Jabiru. She's already built one in West Timor. She's very well known, apparently.' Anna comes to a stop. She doesn't want to tell Sandra that she and Alex had talked about branching out into the tourism industry in Jabiru, should there be an opportunity. Nor does she want to admit that they talked about Safira's aunt being a golden opportunity, given their in depth knowledge of the situation on the ground in Kakadu National Park, and the relationship with Jabiru. '

'Thank you Anna.' Sandra does not want to hear any more. She quickly moves the conversation in another direction. 'What about Alex's personal relationships? What do we know about his friends, girlfriends, boyfriends? Anyone we should talk to here in Canberra?'

It's Anna who continues, as they are friends outside the office. 'He broke up with his girlfriend about six months ago. As far as I know, he does not have a new close relationship. I think that is why he took up surfing. He had time on his hands on the weekends.'

Sandra brings the talk to a close, and beats a hasty retreat to her office, to send the write-up of what she's just learnt to the management team. She can't think of a way to put a good spin on Alex's probable motive for contacting Bob Wilson. She infers it was an attempt to find out what is going on with Jabiru. Although it still doesn't make sense.

What would Bob Wilson know about a deal between an Indonesian tourism developer and the half billion dollar plan drawn up between the Jabiru Aboriginal Corporation and the NT Government? OK he probably does know what is going on, but he is not going to tell a public servant from Parks Canberra anything, that's for sure. She decides she is not going to gild the lily, she'll write an unvarnished account, recommending no further action, now the young man is dead. Nonetheless she'll make an attempt to find out as much as she can about a tourism business deal with the NT government. She steps out of her office, goes over to Anna's desk and says, 'Anna, if you could spare me a couple of minutes?'

CHAPTER 30

Jen gets an email from Interpol containing the report she's requested about Barry Perron.

She goes over to the coffee machine, to top up on her morning coffee, before returning to her desk, and pulling up the report on the big screen of her desktop monitor.

The subject of this report is an individual of Australian origin believed to be involved in the international trafficking of illicit drugs. The subject left Australia in the early 2000s and has since been reportedly engaged in the importation of narcotics from Thailand to Switzerland. The subject's identity and activities have come under scrutiny due to alleged involvement in drug smuggling operations spanning over a decade.

Investigations by law enforcement agencies from multiple countries have indicated the subject was suspected of involvement in smuggling quantities of narcotics, primarily opioids and synthetic drugs, from Thailand to Switzerland. It is believed that the subject has been utilising various smuggling techniques and networks to facilitate the transport of these illegal substances.

In 2001, the subject departed from Australia for Thailand. Subject has connections in Thailand. From 2010 to 2017, he is suspected of being involved in drug trafficking operations, orchestrated from Northern Italy, coordinating the shipment of narcotics from

Thailand to Switzerland using routes and methods to avoid detection. Subject is believed to have been involved in transporting substances from Lombardy via the Aosta Valley in northern Italy and the Grand Saint Bernard Pass into Switzerland.

In 2015, he came to the attention of Swiss Federal Police, Fedpol. They suspected him of being a courier, picking up from international shipping containers with narcotics concealed within legitimate cargo being transported into the Port of Genoa. Subject is known to have utilised encrypted communication channels and secure online platforms to coordinate with accomplices and distributors. Believed to be a minor part of an Italian operation with deep resources, who use techniques such as frequently changing travel routes and aliases to complicate tracking efforts, sophisticated money laundering techniques including blockchain, casinos, and illegal cash smuggling to banking havens.

Fedpol was unable to find evidence. It is noted that he has family in Switzerland in the canton of Valais, close to the Italian border, and connections in Aosta on the Italian side. It is believed that he made use of the legitimate business of bus tours from Switzerland to Italy to return cash over the border.

Subject has acquired an Italian passport, according to Italian National Police, due to his father having been born in Aosta. Plans were underway for a sting operation in Switzerland, abandoned when the subject disappeared from view. Subsequent investigations reveal that he entered Indonesia, remaining in Bali.

Interpol does not endorse or make any representation regarding the accuracy or reliability of information contained in this report.'

From what she has already learnt, Jen is not really surprised that Barry Perron disappeared when the police were getting close. He seems to be a slippery character. She wonders whether he had a tip off from the Italian drugs network. She decides to give Rachel a call and let her know what she's discovered. 'Rachel? We've got a response back from Interpol, and it makes interesting reading.' She proceeds to brief her colleague on the contents of the report.

Rachel listens intently. 'Thanks Jen, that's really helpful. We are going to take a close look at Greg Crowther and his relationship with Barry Perron in relation to the circumstances around a drowning at Gunlom Falls. It may or may not be related to Crowther and Perron. I'll do some more digging. Crowther was one of the last contacts made by the deceased in the investigation. We haven't ruled out foul play yet. If there are any other indications that Crowther and Perron are connected, then I'll apply for authorisation to get call records for Alex Petersen, the deceased. We have a witness who saw him on the phone to someone a couple of hours before his death. His work says no-one was expecting or received a call from Petersen at that time, and his parents had not heard from him either. Leave this with me, if we get anything that could be connected to your investigation, I'll get back to you as soon as possible. '

'OK, thanks Rachel. Look forward to hearing from you.'

CHAPTER 31

'So what have we got to go on with, Rachel?' On Tuesday morning, Isabelle walks into the common desk area, and addresses the young woman sitting at a desk with a coffee and a doughnut, and her laptop open in front of her.

'Well. I've asked Greg Crowther to come into the station voluntarily. He's confirmed for this afternoon, around three. ' Rachel continues with an update on the vehicles she's been researching. 'Barry Perron owns a dark maroon 2017 Holden Calais that he bought new from a dealer in Darwin. Greg owns a 1997 Honda Civic silver hatchback that has been registered to him for the past two years. Alex Petersen hired a Ford Everest at Darwin Airport for one week on a Parks Australia credit card. It's a dark grey colour, and it's still standing in the parking space outside the serviced studio where he was staying. I've organised for forensics to go over Alex's car this afternoon. They've got a backlog, but have said they can give us fingerprints this morning. Europcar says he only drove in and around Darwin. They said he definitely did not drive to Gunlom Falls. '

'So what vehicle did Bob say he saw Alex in? Wasn't it a Holden Calais?'

'Isabelle, I think you're right. Such as the vehicle owned by Mr Barry Perron. First let's find out what we can from Greg Crowther.'

Greg is a little apprehensive as he enters Darwin City Police, giving his name at the desk. After a couple of minutes' wait, he is ushered into an interview room where two women are sitting, waiting for him.

'Mr Crowther, thank you for coming in to see us. I don't know if you've heard, but Alex Petersen, whom I believe is a friend of yours, is deceased. He drowned at Gunlom Falls last Wednesday evening. Just as a matter of routine, we are tracking all of Alex's movements since he arrived from Canberra last Friday. We understand that you were with him at the Fiddler on Friday night. I wonder if you can let us know how Alex seemed to you, did he talk to you about going to Gunlom Falls for any reason?' Rachel is pleasant and encouraging.

'Dead!' Greg sounds and looks shocked. Rachel can't help wondering if he's a little too shocked, perhaps overacting, then dismisses the thought as unkind, given the circumstances.

'Yes, I'm sorry, were you close to Alex? His parents have known for a couple of days. I don't know whether you are in contact with Mr and Mrs Petersen?' Rachel continues.

'No, we're not that close. I know Mr and Mrs Petersen, but I haven't seen them since we left school. We haven't been in contact much since. How did he die?' Greg sounds concerned and serious. 'It's just a bit of a shock. We just met up by chance at the Fiddler.'

'We are still investigating the circumstances around his death. The cause of death was drowning. We're just wondering if he discussed his intention to visit Gunlom Falls while he was here.' Rachel is conducting the interview, and Isabelle sits at the table, writing. 'So did he tell you why he was here?'

'Oh yes. He said he was here for work. Something about attending a meeting with Parks Darwin. ' Greg looks helpful.

'Nothing about Gunlom Falls?' Rachel tries again.

'No.' Greg is now looking careful.

'So what did you talk about? I understand you were deep in conversation with Alex for nearly an hour, according to witnesses.'

Greg now sounds careful. ' We just talked about old times. '

Rachel hones in with a question. ' I believe you know Mr Barry Perron?'

'Yes, what of it?' Greg is definitely wary now.

'Did Alex know Mr Perron?' Rachel wonders if she's gone too far, hoping she's not spoiling the pitch for the ongoing DOCS investigation. 'Did he mention a meeting with Mr Perron at all'

'No.' Greg is clearly rattled.

Isabelle looks up, and takes over the conversation. 'Thank you Mr Crowther, we'll let you know if we need to talk to you again. Thanks for your cooperation. Can you find your way out?'

'Yes, I think so.' Greg stands up, leaves the room, and heads for the foyer and the front door of the station.

Rachel heads for Isabelle's office. 'So what did you think?' They exchange glances. 'I think he was nervous.'

'Do we have enough to call in Barry Perron for an interview?'

'Mr Slippery?' Rachel looks thoughtful. 'Not unless we can get a positive identification that it was Perron's car out at Gunlom Falls when Bob Wilson met him there, or that it was Perron's car that was driven at speed later on past the astronomers at Kambolgie Campground. We can't really call him in on Bob Wilson thinking it was a Holden Calais that Alex was driving, or can we?'

'No. We'll have to liaise with DOCS if they've got him under surveillance.' Isabelle is firm.

'OK. So where do we go from here? '

'I don't know Rachel. Let me think about it. I've got a call scheduled with Alex's boss at Parks Australia later today. I'll see if Sandra has anything else for us. Otherwise I don't think we can take it any further.'

Signing into the Teams Meeting invitation from Sandra Murphy at the appointed time, Isabelle is immediately guested into the online screen, where she can see a middle aged woman seated at a desk with a map of Australia's national parks on the wall behind her. 'Isabelle, nice to meet you. We've emailed, but it's always nice to put a face to the name.'

'Likewise, Sandra. Happy to talk. Have you any more light to shed on Alex's trip to Gunlom Falls?' Niceties over, Isabelle cuts to the chase.

'Indeed I have, Isabelle. This is in confidence, and I am not going to put it in writing, and I would prefer not to have to discuss this outside of this meeting. Parks Australia will of course make every attempt to assist the NT Police in this matter. However I've discovered there are a couple of delicate matters that may involve international affairs, so I would appreciate it if you could be circumspect.' Sandra hesitates for a moment.

Isabelle decides to reassure her. 'Actually we are at a dead end, Sandra. We'd be grateful if you can shed any light.'

'Well it's about Alex's motive for inviting Bob Wilson to Gunlom Falls. I had a talk to one of his colleagues here in Canberra. Apparently he's become friendly with an Indonesian embassy staff member, the niece of a hotelier, a woman called Dewi Pratama who runs a hotel chain, based in Bali. The Indonesian staffer told Alex her aunt has been approached by a Malaysian company called Persekitaran, about a partnership deal with the NT Government and the Jabiru Corporation to build a resort suitable for wealthy tourists. Apparently she's already done this in West Timor, very successfully. Now this is not about her company. They are very reputable, very environmentally friendly, their developments are sustainable, and they run renewable energy projects to power their hotels, as well as a couple of waste-to-energy facilities in Bali.

It's Persekitaran. They are well known for being involved in a lot of, shall we say, shady developments. Illegal forestry in Malaysia and New Guinea, Africa, and Indonesia. And it's well known that they have close links to China. Their business methods are to find a local person in the place where they want to operate, set up a joint directorship with a local company, and then use the local company as the developer. They choose people they think they can easily manipulate. They have started branching out into tourism. So it seems that Persekitaran has engaged with a company based in Darwin called Top End Tourism. They would be the ones who would engage the hotel developer for the proposed high end resort.

Now I am not saying that Alex has anything to do with Persekitaran. I'm led to believe that he put out some feelers about the proposed tourist development at Jabiru. And he got a response back from an old school friend in Darwin, asking Alex if he'd like to meet the director of a company involved in the deal, and that there might be some business in it for both of them. Really Isabelle, that's all I've managed to find out.'

'His colleague did not happen to mention the name of the old school friend, by any chance?' Isabelle asks.

'No, that's all she, err Alex's colleague knew. She said Alex had only referred to him as Greg.' Sandra sits back, unsure if she's said enough, or too much. She's already relayed this information to the management team, and they directed her to get back to the NT Police.

'Sandra, I'll be discretion itself. Thank you. That is really helpful. I'll let you know if we can shed further light on Alex's death.'

'Please do. Thanks Isabelle. Thanks for your time.'

By the time Sandra closes the meeting, Isabelle is already on ASIC, the Australian company regulator, doing a company search for Top End Tourism. 'Bingo,' she says aloud, scribbling some details on a yellow post it sticker. Old habits die hard. She heads out to the open desk area.

'Rachel, I've just spoken to Sandra Murphy from Parks in Canberra. Alex confided in one of his colleagues. He was in contact with an old school friend with whom he was discussing a business deal. She gave me the name of a Darwin company, supposedly involved in a hotel development project with the NT Government, the Jabiru Corporation and a shady Malaysian outfit called Persekitaran. Alex was to be put in contact with this company. '

'Did she name names?' Rachel asks.

Isabelle reads from her scribbled notes. 'The company name is ‘Top End Tourism’ and there are two directors, Lim Rajendran, whose address is given as Smith Street, Darwin City, and, drum roll, Barry Perron, Lambell Terrace, Larrekeyah. And Sandra only knew the first name of the old school friend. Greg.'

Rachel gets up from her desk and comes round to read the note. 'That’s amazing, Isabelle. Who do we interview first, Perron or Crowther?'

'Let’s get Bob Wilson back in first. We can drive to Jabiru first thing tomorrow morning. See what he knows about the tourist development, and tell him we know that Alex Petersen was looking for a business deal. I don’t think he had anything to do with Alex’s death, but I do think he knows more than he’s let on. Then we go after Perron, pulling in Crowther at the same time, with as many ducks lined up as we can. See if we can find out exactly what happened at Gunlom Falls.'

'Sounds like a plan.' Rachel secretly thinks that she’s done enough driving this week, but she knows that Isabelle’s right. They’ll get more from Bob Wilson with a softly softly approach on his turf, than by doing an online interview, or getting someone from Jabiru to drive him to Darwin. 'I’ll get on to Jabiru and get them to line up Bob Wilson.'

'And you can get back to your friend in Drugs and Organised Crime to tell her about Perron's company and links with Persekitaran. ' Isabelle retreats to her office.

CHAPTER 32

The following morning, after an uneventful drive, the two policewomen once again pull up in front of the Jabiru police station, with coffees and bacon rolls acquired at the Kakadu Crocodile on their way through.

'Interview commenced at 10.20, in the room Detective Senior Sergeant Isabelle Felham, and Detective Sergeant Rachel Bonham with Mr Robert Wilson in connection with enquiries into the death of Alex Petersen at Gunlom Falls. Thanks for coming in Mr Wilson. We've been talking to Alex's boss in Canberra. We now know that Alex may have invited you to Gunlom Falls, not as an employee of Parks Australia, but in connection with a private business venture. Did he give you any indication that he might be representing a developer called Top End Tourism?' It's Rachel who is doing the talking.

'No. He didn't say anything about Top End Tourism.' Bob's response is quick.

'So what did he say?' Isabelle weighs in. 'We know he was probably involved with a shady dealer who wants a piece of the action in the redevelopment of Jabiru by the NT Government and the Jabiru Corporation. He must have told you something more than asking you to show him the stairs. Parks Australia has already been out to talk with Jawoyn traditional owners several times. He had absolutely no authority from Parks to talk to you about the new stairs and fixing the damage. So what did he tell you?'

Bob is silent for half a minute, clearly composing his thoughts. 'He did say something. After I showed him the stairs, and told him what's wrong with them, he asked me if we were interested in redevelopment. He said that there was interest in doing the right thing at Gunlom. He said he was approached by some people involved in the redevelopment of Jabiru who would pay to move the stairs, and make a new track up to the falls in a culturally appropriate place. And that as part of the deal, the company would pay for improved parking facilities.

So I asked him what they wanted for that. At first he just said that they were helping Parks Australia to upgrade the infrastructure. I know Parks doesn't work that way, and anyway, any upgrades would come out of Federal upgrade funds, and it would be discussed in formal meetings with the Kakadu Board. So I said I didn't believe that. And I asked him again what he wanted. He said that in return for Jawoyn support for the development at Jabiru, the company would see that upgrades to Gunlom, Boulder and Kambolgie parking and facilities were part of the deal. And that there might be something in it for the Jawoyn. '

Isabelle and Rachel were looking at him encouragingly. 'So what did you say?' Rachel asks.

'I said no. I don't know what kind of mug he took me for. Anything that gets done in Kakadu goes through at least three layers of approval. It has to be sanctioned by all the traditional owners, not just the Jawoyn. It gets discussed at board meetings, it has to get run by the NT Government and then it has to get buy-in from Parks. And they are the slowest of all. They usually delay any decision for months, sometimes years.

We can't get funding to fill in potholes because Parks sits on the money doled out by the Federal government. And the Jabiru Corporation is a private company. They have put up a business proposal for the redevelopment, through a deal done with the feds, the NT government and the mining company, ERA. Believe me, anything done at Jabiru has to climb up the greasy pole of dealing with that lot, before Jabiru Corporation gets a say. '

'So what happened then?' It is Isabelle who speaks, though both Isabelle and Rachel are leaning forward in their seats.

'I told him to come back when the people he's talking to have a contract in place to do the development. I was polite. But he could see I meant it. So we said goodbye and walked back to the carpark. I got in my car, then I drove off. By that time he'd got back in his car, and was sitting in the passenger seat, and he had his mobile phone in his hand. '

'Why didn't you tell us this last time we interviewed you?' Rachel is curious about whether it was reticence, or circumspection.

'You didn't ask me what he said. It happened just as I told you last time. We walked around, and I showed him the stairs and pointed out that it was too close to one of our sites, and that the Jawoyn are still working out what we want done. That's when he said all that stuff. He might have thought that he could offer us money and a change of location for the stairs. But that is not how we work, how we think. We want Gunlom Falls to remain a culturally sensitive place. And new stairs and a carpark could be considered, it'd have to be on our terms. OK the sealed road is tempting. But I could see he had no idea what kind of bureaucracy everything goes through to get anything done around here. I thought he was trying it on, dreaming of being a big shot developer. I did not think he was talking for Parks in Canberra, because they don't work that way.

Do you still want proof of what time I filled in my travel diary? I got an email from IT. I can forward it to you.'

'No, that's all right, Mr Wilson. No need for that now. Thanks for coming in.'

On the way back to Darwin, Isabelle addresses Rachel. 'I think I just learnt something. And that is that the Jabiru redevelopment is as dirty as big deals usually are around here. Maybe I'll go on that Cultural Sensitivity course.'

'Good thinking Isabelle. And I read that ERA wants to reopen the mine, and that they're lobbying the feds to allow them to start mining uranium again. So now what?'

'We interview Perron and Crowther, and get forensics to go over Perron's vehicle for Alex's fingerprints ASAP. I wonder why he sat in the passenger seat? Where were Perron and Crowther while he was at Gunlom Falls? If they were there, Bob would have seen them. Nowhere to hide out there, and it's the end of the road.'

'They could have been up at the Falls. Bob said they didn't go up to the top, just stayed near the bottom of the track.'

'Good thinking. We need to do that as soon as

possible. We mentioned we knew about Crowther's relationship with Perron yesterday. Unless I miss my guess, he'll tell him as soon as he sees him that we are interested in him.' Isabelle takes out her phone and puts in a call to the Scene of Crime forensics unit.

The following day, promptly at 10 am, Barry Perron is in for an interview in the Darwin City Police Station. When confronted with the supposition that fingerprints in his Holden Calais belonged to Alex Petersen, now deceased, his answer is succinct. 'Yes, I am a director of the Top End Tourism company. And no, I have no idea what Alex Petersen was doing in my vehicle. I loan it out from time to time, to a young man who sailed a boat from Bali for me. When he wants to go off the sealed bitumen onto the dirt roads, he sometimes asks if he can borrow my vehicle, as his car isn't up to the backroads.'

'Did you loan it to Greg Crowther last Wednesday?' Rachel senses she's not going to get much out of this interview, so she focuses on specifics.

'Yes. Now if there's nothing else, I am a busy man.' Barry Perron rises from his chair.

'One more thing. As we have not ruled out a suspicious death in the case of Alex Peterson, and we're following up on a report that he was sighted in a vehicle whose description matches your vehicle, we'll be applying for approval to run fingerprint and DNA checks on your vehicle. Just for the purposes of elimination. That's all Mr Perron. Interview terminated at 10.15.' Isabelle swipes to close the recording facility on her phone.

Barry Perron walks out of the room, thinking that he may have to change his plans.

In the recap, Isabelle and Rachel agree that Perron is indeed hard to pin down, and that they'd have to get a warrant to see his phone and email records to establish contact with Alex Petersen, before they can question him further, even if they find his fingerprints and DNA in Perron's car. 'We're dealing with a consummate professional. He'd never expose himself or take risks unnecessarily. I bet that all the contacts were made through Greg Crowther.'

'Yes. He'll be here in half an hour.' says Rachel, also heading for the door. 'Just time for a coffee and a blueberry muffin?'

'Double latte and a cinnamon and apple doughnut.' Isabelle does not look up from her laptop, where she is furiously writing up the latest interview.

'Speaking of contacts, what did forensics learn from Alex's mobile phone? We know that there's no reception out at Kakadu, but at least we can see what contacts Alex had with Crowther. I've already asked Parks to provide us with a list of his work mobile calls and callers,' Isabelle asks.

Rachel looks blank. 'They found his mobile in his hire car. I didn't see anything out of the ordinary in the report. A couple of calls to his parents. I would think they'd have reported anything out of the ordinary. '

'I ran it past the Super yesterday, and he's given approval to take Crowther's fingerprints and to check whether there are traces of Alex in Perron's vehicle. We can get the check done on Perron's vehicle tomorrow.'

Isabelle turns and walks towards her office. 'We'll hint to Crowther that Alex was seen in Perron's vehicle. We might want to interview Perron again, after talking to Crowther. Can you book an interview room for three?'

'Already done, Isabelle.' Rachel has already turned back to her laptop.

'We'll take Crowther's fingerprints at the start of the interview. Then we'll see what he has to say about where he was when Alex was with Bob Wilson. Then if there is a match, we'll request a warrant for Crowther's mobile phone records, given that forensics has not ruled out Alex's death as suspicious. '

In the interview room, after the initial formalities, it's Rachel who asks. 'Before we start, can we take your fingerprints?'

'Why do you want to take my fingerprints?' Greg suddenly looks alert.

'Just routine, Mr Crowther. We can require fingerprints in a case like this, where the cause of death is not fully determined. We have approval from a Superintendent of Police. Before fingerprinting, can we see your driving licence? We just need an identity document with your photo on it. Along with the fingerprints, your details are stored on a secure digital database with your name, date of birth, and driver's licence.'

Isabelle makes a call, and a police officer appears to escort Greg to the fingerprinting room, with state of the art electronic equipment for capturing clear and accurate fingerprints. 'If you can just clean your hands with this sanitiser, then dry them. This is just to ensure we get a good quality impression. ' It's the police officer in attendance operating the digital scanner who issues the instructions.

Meanwhile in the interview room, Isabelle and Rachel confer. 'I think he looked uncomfortable, especially when we asked for his fingerprints. As soon as we have the prints back from Perron's vehicle, we can get an immediate automated comparison with AFIS.'

'So how does that work? I've only used the manual system.' Isabelle asks. 'Although I am not sure I want to know.'

Rachel laughs. 'It's OK, I studied computer science. The system uses algorithms to analyse the ridges and specific points to compare whether there is a potential match.'

'Well at least it's quick. So we're expecting to get a match for both Alex Petersen's and Greg Crowther's fingerprints with the fingerprints in Perron's vehicle. So we'll confirm the suspicion that the vehicle Bob Wilson saw Alex in at Gunlom Falls was Perron's. So when we get the prints back from the Forensics Unit tomorrow, we'll know immediately?' Isabelle is used to a fairly lengthy fingerprint matching process.

'Yes. We'll know tomorrow morning if Crowther has driven Perron's vehicle. Though as Perron has admitted he loans his vehicle to Crowther, that's probably a dead end. And if Alex's prints are in the vehicle, did both Crowther and Perron drive Alex to Gunlom Falls? If so, why?'

'Well, we'll soon know. Let's ask Crowther directly. And find out exactly what role he plays in Top End

Tourism.'

Isabelle lowers her voice as the door opens, and Greg Crowther has been escorted back to the interview room. 'We know you were talking to the deceased, Alex Petersen, about a potential business deal with Top End Tourism. One of the company directors is Barry Perron. Were you at Gunlom Falls last Wednesday with Alex Peterson, Greg?'

'No.' The denial sounds like a knee jerk reaction to Isabelle and Rachel.

'So how do you account for Alex Petersen being in Barry Perron's car at the Falls? Mr Perron has already told us that he loaned you his four wheel drive to go on a backroads trip last Wednesday.' Rachel follows up.

Greg fails to spot the trap that the police have laid for him. They are not sure that it was Barry's car at Gunlom Falls. He makes a hurried mental calculation that Alex's fingerprints will be in Barry's car, and if they've taken his fingerprints, they'll know he was in the car too. He quickly decides on a new tack. 'Oh, you mean last Wednesday? Yes, Alex said he had to go to Gunlom Falls for a meeting with someone from the Kakadu Board. I decided to go too, to see what it's like now. We used to go swimming there on weekends when we were at school.'

'Mr Crowther, before you dig yourself a big hole that you can't get out of, I suggest you tell us the truth.' It's Isabelle who responds. 'You'll save yourself a lot of trouble. We have already asked for a warrant to collect DNA and fingerprints from Mr Perron's vehicle.' She doesn't add that she's not sure they have enough evidence to have the request approved.

Greg has clearly decided that a show of frankness is his best bet. 'Well yes we did go to Gunlom Falls. Alex said he had an important work meeting. He said he just wanted a lift to the Falls, and that he'd get a lift back to Jabiru, where he had a meeting the next day. '

Rachel weighs in. They both sense that Greg is feeling the pressure.'So you dropped him at the Falls, and went back to Darwin. '

'Yes.' By this time Greg Crowther is clearly thinking very hard

'Was anyone with you?'

'No.'

'So where were you? A witness saw Alex Petersen get into the maroon Holden Calais at Gunlom Falls, Barry Perron's maroon Holden Calais, and there was no sign of anyone else around. No sign of you.' Rachel is still reconstructing the scene. 'Can I jog your memory? Alex did not get a lift to Jabiru. Last time anyone saw him alive, he was in Perron's car. Alone. And the next witness statement we have is that he was up at the pools, in the water, dead. So he must have walked up the track. There's only one track. So where were you? Were you up at the pools? '

Isabelle decides this is the moment when they try to get Greg to tell them what really happened. 'Did he join you? Did you sit down on the rocks above that incredible view looking out at the beautiful Kakadu sunset? Did he tell you that the plan to get a piece of the action with the tourism development at Jabiru with Top End Tourism had come unstuck at the first hurdle? Is that what happened Greg? You were both very disappointed, I suppose. You were counting on Alex being able to use his influence at Parks Australia to get the buy-in from the Jawoyn people. Because Perron needs indigenous groups to lend support to a potential Jabiru development deal with Top End Tourism. '

But Greg wasn't born yesterday. He has thought up a story, and he's going to stick to it. 'No, that's not what happened at all.' He coughs several times, then asks for water to buy some time. When Rachel returns with a glass of the best Darwin tap, he has composed himself. 'I didn't see Alex at all after I left him in the carpark at Gunlom. I did go up the track. I'd come all this way, and I wanted to see the infinity pool and the view. By the time I got back to the car park, he wasn't there. So I presumed he got the lift he wanted to Jabiru. Said he had to go there for a board meeting, and didn't feel like going all the way back to Darwin. Said he'd stay overnight in Jabiru, then get a lift back the next day with one of his colleagues from Parks. '

'Alex Petersen's body was found at Gunlom Falls. He suffered a blow to his head, either someone hit him or he fell on a rock at the water's edge, ended up in the water unconscious, and drowned. So how come you didn't see Alex up there. There is only one track. You couldn't have missed him. You'd have to pass him coming up while you were going down. What aren't you telling us, Mr Crowther?' Isabelle asks the leading question this time.

Greg knows exactly what he is being asked, but he can see daylight at the end of the tunnel. He'd always been good at talking himself out of trouble. 'Oh, I didn't go up the track, I climbed over the barriers of the new stairs Parks built. I went up and down that way. If Alex was on the track, I wouldn't have seen him. It's really sad. If I had seen him, he might not have had the accident.'

'So what did you do after you walked down from Gunlom Falls to Perron's car? Couldn't have been much after six. Mr Crowther, you were in the middle of nowhere, with no facilities, no mobile phone signals in the Kakadu National Park, no one around.'

Isabelle and Rachel have only one card to play, and they are hoping it's an ace. 'So what time did you drive back to Darwin, Mr Crowther?' Rachel asks the obvious question.

Greg is metaphorically thinking on his feet again. He reviews what he knows about the location of fixed speed and red light cameras operating in Darwin. He's pretty sure there is permanent number plate recognition software that the police can access, on the northbound Stuart Highway approach, and decides on what he is going to say. 'I didn't drive back until about midnight.'

There is an incredulous note in Isabelle's voice.' You stayed at the carpark for six hours, until midnight, doing nothing, with no mobile phone signal and no internet, before you decided to drive back to Darwin?

This time Greg feels he has the situation under control. 'I had a sleep in the back of the car. I was really tired after the long drive out. And I did not want to drive back and fall asleep at the wheel. You know the police are always running campaigns 'don't drive when you're tired, it can kill you'. Well I didn't. I had a sleep instead. When I woke up, I drove back along Gumbat Road then onto the highway. I got back about four in the morning. '

'Interview terminated at 3.50 pm. You can go now, Mr Crowther. We've finished. For now.' As soon as the door closes, Isabelle and Rachel sit in silence for a couple of minutes. 'I don't see what else we can do. He's lying, of course. The problem is that there is no way we can track what happened to Alex. No mobile phone reception. No witnesses. Nobody around for at least ten kilometres. Nothing, nada. ' It's Rachel who breaks the silence. '

'Yes, he's lying, and there's not a damned thing we can do about it.' Isabelle concurs. 'Either Alex had an accident, fell, hit his head, and Crowther just left him in the water to drown. Or they had a bust up because Alex told him there was no way he could get a deal happening for Top End Tourism, once Bob Wilson had turned down the offer of cash to the Jawoyn in return for their support. I suppose he realised that he wasn't going to be able to deal with the incredible bureaucracy of two governments, one multinational mining company, an aboriginal corporation and a board that not only involved several indigenous groups, but also his own employer, Parks Australia. He must have realised that there was no deal to be done that he could be involved with. He probably realised that Persekitaran and Top End Tourism would be involved in out and out corruption to get the redevelopment contract for Jabiru, by greasing all the palms involved. He couldn't afford to have it known he was associated with anything shady because he'd lose his job with Parks Australia.'

'I agree Isabelle.' Rachel takes up the thread. 'Bob Wilson put it straight that there was no deal to be done. Top End Tourism was not his dream business opportunity to strike out as a property developer. And he'd be lucky if what he said to Wilson did not get back to Parks. I could imagine that there would be a big argument between them. Crowther is obviously in deep with Perron, and Perron is calling the shots. I suppose he did not want to go back to Perron and tell him that when he said he could get a deal done through an old mate who is high up in Parks Australia, he was just boasting.'

'Given that there was no-one around for miles, I don't think we can go any further. The trail's gone cold. We'll never know exactly what happened to Alex Petersen, and that's really sad.' Isabelle powers down her laptop, and rises from the table.' I'll write this up. I'll send a copy of the report to the Coroner with the conclusion that we are unable to determine that anyone else was involved in his death. And I'll give Sandra in Parks a call, and privately tell her what we heard from Bob Wilson, and that we have our suspicions that his death was not entirely accidental, but that we cannot prove anything.'

'OK. I don't like leaving it at that, but I can't think of anything else we can do either. Except to let DOCS know that both Perron and Crowther are dodgy as hell, and hope that they get tripped up in one of Perron's illegal drug activities.'

'I am going to confer with the Super, and ask him to write a report to both the NT Minister and the Federal Minister involved with the Jabiru redevelopment, and let them know what we suspect about Top End Tourism. Hopefully if we write to both levels of government, that will eliminate them from being part of any Jabiru tourist development. Because it brings Persekitaran's integrity into question. Neither government could be seen to be involved with something that has a nasty smell, and nor could Rio Tinto and ERA afford to be in bed with them when there is an NT Police Report about Top End Tourism, Persekitaran's local partner, with Perron as a director, given that he's under surveillance on suspicion of drug running. ' Isabelle heads for the door, as usual throwing a remark over her shoulder as she moves into the corridor. '

Rachel follows. 'Best we can do, Isabelle.'

CHAPTER 33

'The AUKUS military deal is an agreement between Australia, the United States, and the United Kingdom to collaborate on defence and security matters in the Indo-Pacific region. It was announced in September 2021, and includes a range of initiatives, including the development of nuclear-powered submarines for the Australian navy.'

Graham Saunders is reading from an autocue on the screen in front of him, strategically placed so it seems as though he is directly addressing the camera in the ABC studio, and the audience at home.

'The AUKUS deal represents a significant shift in the balance of power in the Indo-Pacific region. It signals a growing alignment of military powers. The focus of the agreement is on enhancing Australia's naval capabilities, particularly in response to the increasing territorial assertions of China in the region.

One of the key elements of the AUKUS deal is the development of the submarines, a significant upgrade from the current conventional diesel-electric submarines. These vessels are quieter, faster, and have a longer range than their conventional counterparts, making them better suited to operating in the vast distances and challenging conditions of the Indo-Pacific.'

Graham's voice over is now emerging from footage showing the existing Australian vessels, and mock ups of the new proposed fleet of eight submarines.

'The AUKUS deal has attracted criticism from some quarters, particularly from France, who had previously been promised a deal to supply conventional submarines to Australia. The French government was reportedly blindsided by the announcement of the AUKUS deal and accused Australia of acting in bad faith.'

The TV footage shows an angry French president, addressing a news conference after being informed by text that the deal is off.

'Despite the controversy surrounding the AUKUS deal, according to the government, it represents a significant step forward for Australia's defence capabilities. It is likely to have a major impact on the strategic balance in the region for years to come.'

'Like hell', Graham thinks to himself. 'What's the point in angering China, a dragon with a thorn in its foot, when Australia wants to resume its lucrative trade deals? And of course the development of new drone technology could make the submarines obsolete within a decade, well before they are delivered.'

In the ABC studio in Parliament House, Canberra, the camera pans to a suited politician seated carefully at a table. Graham addresses the Defence Minister.

'Minister, what is the role of nuclear-powered submarines in Australia's defence posture?'

Blair Maddox faces the camera with a gaze of blue eyed simplicity.

'The AUKUS deal primarily involves the development of nuclear-powered submarines, which are different from nuclear weapons. They use a nuclear reactor to generate electricity, powering the submarine's engines and other systems. The nuclear fuel used in these reactors is highly enriched uranium. '

Graham uses a pause to ask another question. 'Under the agreement, Australia will be responsible for storing high-level nuclear waste from the decommissioned reactors. And that is no easy feat. Would you care to comment on that, Minister?'

'Well,' the Minister pauses, perhaps to consider his response to a not entirely unexpected question. 'It is likely that the waste would be stored onboard the submarines and then disposed of according to established protocols. These submarines have been used by other countries, such as the United States, for decades, and there are established procedures for managing waste.'

Graham continues on the line of questioning. 'Once removed and decommissioned, a complicated and hugely expensive process, there remains one big question – where will this waste be stored? ' Graham knows where he wants to lead next, but is not sure he'll get a definitive answer.

The minister deftly sidesteps the question. 'AUKUS does not involve the deployment of nuclear weapons in the region. Australia, along with the US and the UK, are signatories to the Treaty on the Non-Proliferation of Nuclear Weapons. We are committed to using the technology only for peaceful purposes, as is required by the International Atomic Energy Agency.'

'Minister, the submarine reactors are fuelled by

highly enriched uranium-235.' Graham steps in with an observation, trying to pull the interview back to where he wants it to go, a statement on where the nuclear waste will be stored. 'The government has not given any details, other than it is to be stored on land that is either owned or acquired by the Defence Department. Would you care to comment on where the spent fuel will be stored?'

'No, I have no further detail at this time, Graham.'

The visual crosses to Maralinga, a deep red earth, sparsely vegetated landscape, and a vast blue sky, where Graham is speaking in measured tones, directly to the camera, standing beside a sign that carries the nuclear warning symbol, a trefoil around a small central circle representing radiation from an atom, and the words - Warning, Buried Radioactive Materials.

'I am standing in Maralinga country. In the 1950s and 60s, the UK conducted seven nuclear tests at the Maralinga site in South Australia, part of the Woomera Prohibited Area, north west of Adelaide. Aboriginal people have occupied the Maralinga Tjarutja lands for tens of thousands of years. This country holds cultural, social, economic and spiritual significance for the people. In January 1985, native title was granted over some of the land, even though a Royal Commission identified there was significant residual radiation contamination at some sites.'

A voice over takes over the commentary from Graham, while footage from the station's archives shows various British and Australian personnel at the Maralinga site.

'The nuclear weapons tests resulted in widespread contamination of the area, and forced the Maralinga Tjarutja people to be involuntarily relocated. The testing program, known as the British Nuclear Tests in Australia, was conducted with the support of the British government, which used the Maralinga site to test nuclear weapons as part of its Cold War defence strategy. The Maralinga Tjarutja people were not adequately consulted or compensated for the loss of their traditional lands, the contamination and the environmental damage.

The Maralinga Tjarutja Trust was established to manage their land and resources, and the people are now involved in a range of cultural and community activities. Their experience with the British Nuclear Tests in Australia has had a significant ongoing impact on their culture and way of life, an important reminder that the impact of radiation contamination is extremely long lasting.'

The program now switches to historic footage of Woomera Village. 'The Maralinga people have a right to be suspicious. Established in 1947 to test British long-range missiles and other weapons, Woomera Rocket Range was the hub of space activities in Australia from the 1950s to 1980. The Woomera Prohibited Area is now used for weapons testing for the Defence Department. The area encompasses the traditional lands of six Aboriginal groups, including Maralinga Tjarutja and Anangu Pitjantjatjara Yunkunytjatjara. The history of these people and their deep ties to the land date back over tens of thousands of years.

The land contains sites of enduring significance

to Aboriginal people, including stone arrangements associated with traditional ceremony and ritual, rock art sites, ceremonial sites, cultural sites, watercourses, and archaeological sites that show how people lived in and used their environment. Twenty years ago Woomera land was used for a scandal ridden controversial detention centre for refugees, before being handed back to Defence.'

The visuals cross back to the studio, where Graham is standing facing the camera, reading from the auto cue. 'Nuclear submarines, particularly based on UK and US designs, might never need to refuel. They use what's called highly enriched uranium, which enables them to run for the whole life of the submarine, up to thirty years, without being refuelled. Nuclear reactors pose potentially serious risks for people and the environment, particularly if there is an accident at sea.

The key difference between submarine and land-based reactors is motion. A land-based reactor doesn't move. And of course submarines roll from side to side, meaning that there has to be additional measures to control the chain reaction and the thermal power output of the reactor. To make sure that the reactor can always be shut down, despite the motion, there has to be more control rods. This raises the stakes if there is an accident.

Radioactive wastes are produced from manufacturing, maintenance, operation and decommissioning of nuclear submarines. Earlier this week, Defence Minister Blair Maddox reaffirmed that the submarine waste would be stored on Defence Department land.'

The visuals return to the defence minister sitting in the ABC studio in Canberra.

'Minister, we understand the favoured location for high, medium and low level nuclear waste is the Woomera Rocket Range in South Australia.' Graham decides to ask this question first.

'No decision has been made yet, Graham.' The Defence Minister looks uncomfortable, he was not expecting this question.

'Minister, can you confirm that a US Naval Base is to be opened in Darwin, as discovered by the ABC from US budget papers? '

Maddox was not expecting this question either. 'I cannot comment on matters of national security.'

Graham continues. 'Can you comment on a report that the Jabiluka Uranium Mine owned by Rio Tinto subsidiary ERA is to be reopened, despite opposition from traditional owners. This would provide more fuel, for example, for nuclear submarines?'

'Graham, this is completely unsubstantiated. I am sure you understand I cannot comment on matters of significance for Australia's national defence. '

'Have you discussed this with the US Government?' Graham presses home.

'I can only repeat that I cannot comment on matters of strategic significance.' Maddox appears to be thinking hard about how to dodge this line of questioning.

'Yes, I can understand that minister. So here is a question closer to home. Will there be new uranium mining at Jabiru, in the midst of Kakadu National Park, which is already under threat from delisting as a World Heritage site by UNESCO?'

'That is a matter for the Minister for Resources, not for the Defence Minister.'

'So the government has not rejected the possibility. Surely, Minister, this is a matter discussed in cabinet? Can you confirm that the new port facilities planned for Middle Arm will include uranium enrichment facilities powered by gas fracked at Beetaloo Basin?'

'No. I am not confirming any such thing.' Blair Maddox now looks distinctly uncomfortable. 'The port facilities at Middle Arm are part of my government's green initiative for sustainable development.

'Thanks for your time, Minister.' Graham lowers his notes, and attempts to look sincere for the twenty seconds before the program ends, replaced on the national TV network viewer's screens with an infomercial for a comedy program.

CHAPTER 34

Yindi is worried. They are talking around the campfire of how the firestick farming that has allowed her people to manage the grazing of the animals, is changing how much rain falls. 'They have gone too far, burning too much, so that the country is changing too fast. Is this why the big wet season rains were less than previous seasons?' she wonders. 'The land needs the water to grow all the trees again.' Yet there was less abundant growth in the last storm season because the rain did not flood the land as it used to do. The cool season came early, and the billabongs dried up too soon.

She talks it over with her family. Yindi, even though she is young, is a well respected leader, and everyone listens to her knowledge. 'We have to move along to let the trees grow back, or we'll just be left with grasslands and not enough trees to call the storms back next time, and the time after.' Her grandmother, Kuralini, agrees. 'Doesn't do to play with water time. Time might go backwards, and then where would we be? We need the trees and so do the animals. We might be too greedy, hunting too much, clearing too much country. Let's listen to the star beings and see what they have to say.' Kuralini was talking about the ancient practice of reading the stars to see what the clan would do next. To move on or stay. Where to go. Big decisions ahead. Time to listen to the spirit beings. Time for a ceremony. Time to tell the stories and dance the dances. Time to find out. "We do not want to cause trouble for the future,' agrees her brother Waru.

The stars are bright above. The campfire is burning brightly. Yindi's clan is sitting around quietly, and as the night unfolds, the bullroarers are heard above the crackling of the fire. The dancers perform the dance of the water energy meeting the sky energy. In the flickering light while the people sit silently, the voices from the past come murmuring through the atmosphere, telling the story of a time when spirit beings were able to make themselves heard. All the dancers sit down. No-one moves.

The stars are dancing their own dance across the sky, occasionally an asteroid can be seen tracing its path down to Earth. The constellations continue their slow movement towards the direction of sunset, and still nobody says a word. When the fire has nearly died down, Yindi and Waru slowly stand up, and wander to the edge of the ring to collect some more wood. Suddenly a quiet murmur of conversation starts up, and everyone starts talking at the same time. 'I think we have to move', says Yindi and Waru adds 'I wonder what the others think. We have to find water and country where the grass grows right up to the trees, and rain falls properly in the wet season.' 'Yes', responds Waru. 'And now we know what direction to travel, we just have to decide how much water and food we are going to take with us, and who is looking after the elders'. 'Yes, Waru. Big decisions, and not much time if we are going to get going and find our new place before the hot dry season starts. And we don't want to go where people already are. We want friendship. We don't want to take up someone else's water, someone else's hunting grounds. So we have to think carefully where the best places might be, even if we don't know where they are'. 'Yindi, I hope Kuralini is listening to the spirit guides. We can find water OK by following the animals, but we must go where no other clan lives now. And that might be hard to find.' 'I agree. That is what we have to do. Kuralini will know most about that, let's ask her.'

The pair move off to sit with their grandmother, who is tracing out a pattern in the sand. 'I know what you are thinking' she says. 'This is our best move, and we need to go before the next moon, to get where we are going before the next storm season. It's a long way.' The pair study the map the woman has drawn in the sand, with the annotations of the familiar places, easy to read even in the flickering fire light. 'Oh, we have to head for Wariyapu, the sunset rock?'

'Yes.' replies Kuralini.' That way we'll have plenty of water all the way to the new place.'.

'Where is the new place, then, Kuralini?'

'Yindi, I don't know. But I am sure, because it comes from my dream. And my dream comes from the spirit beings who guide our clan. I heard it in the air dancing above the fire.'

Yindi and Waru exchange glances. Their grandmother knows the way! 'Really? Did they tell you to go to Wariyapu?'

There is a twinkle in Kuralini's eye. 'My grandfather told me when I was a little girl that there is a map at Wariyapu showing where there are water sources. It is in the direction back to the seashore where we came from, a long time ago. We'd better go there first.'

Yindi laughs. Should have known her grandmother

would be very practical, as well as listening to inner wisdom!

CHAPTER 35

In preparation for his trip 'Down Under', Fabian revises the conservation statements for the Kakadu National Park, starting with the 1986 report. He quickly discovers that every report lists 'Mining' as the biggest threat. Lately groundwater pollution and invasive alien terrestrial species both appear on the list. Kakadu was designated as a UNESCO World Heritage Site in 1981 due to its cultural and natural significance, stunning landscapes, rich biodiversity, and its cultural importance to the indigenous peoples of the region. In 2021, the conservation outlook was assessed as 'significant concern'. Fabian hopes there is some progress made by the Kakadu Board to allocate funds to repairing damaged infrastructure and to start the rehabilitation process to degraded land.

He decides he must make the strongest possible case for both the natural and cultural values for the largest national park in Australia and one of the largest in the world's tropics, to give it the best chance of retaining its World Heritage status. Kakadu preserves the greatest variety of ecosystems on the Australian continent including extensive areas of savannah woodlands, open forest, floodplains, mangroves, tidal mudflats, coastal areas and monsoon forests. He needs help to determine the chances for maintaining the natural environment, in view of environmental threats.

As well as nature, the heritage value of Kakadu is deep and intense. The first Australians have unique performance arts including music, dance, song, ceremonies, symbols and designs, narratives and poetry. The people who lived on the continent of Australia for over 50,000 years also have deep traditions of scientific, agricultural, technical and ecological knowledge. They have a unique body of knowledge for medical purposes. And huge experience and skills in the sustainable use of flora and fauna.

Fabian is aware that preservation of ancient traditions is priceless. Indigenous people know that there is still a danger that knowledge will be lost forever, knowledge the world cannot afford to lose, to rebalance the weather, climate and biodiversity of the complex systems that keep humanity living in conditions that nurture life. He comes to a decision that there is no time to lose.

Indigenous Australians are still being sidelined in Australian society, because of the colonial declaration of terra nullius, a mistake that has been acknowledged. He decides he will offer all the support he can muster to the Northern Land Council, and the Kakadu Board to maintain the World Heritage status. Perhaps he can be a catalyst for the stakeholders to move in a direction that will enhance the preservation of the oldest continuous civilisation, and the least disturbed nature on Planet Earth. He figures that this ambitious aspiration is worth doing.

The park also has a huge diversity of flora and is one of the least impacted areas of the northern part of the Australian continent. Its spectacular scenery includes landscapes of arresting beauty, with steep rock escarpments extending in a jagged and unbroken line for hundreds of kilometres. He wonders what on Earth can save Kakadu from massive new mining and port industrial developments in the Northern Territory, and the associated huge rise in carbon emissions and the accelerated climate change?

He shares a view that governments are very creative at finding a way to ignore indigenous culture when it suits them. Just as they approve shoddy environment impact assessments that are really damaging interconnected ecosystems, something happening all over the world that stops human evolution in its tracks. There is pressure from housing market lack of supply, so governments are all too happy to approve questionable environmental management plans, resulting in climate risky developments with poor outcomes for people.

He's also doing research on Parks Australia, co-managers of Kakadu, with no resolution to the Gunlom Falls stairs built too close to avoid damage to a sacred site. He wonders whether the federal government reform of national heritage is just a long term vision, kicking the can down the road. He has to write a report on the risks to Kakadu to determine whether it keeps it's world heritage status. Melanie's news that the Australian Government is ignoring cultural sensitivities at Middle Arm, and the rumours that talks are being held with Rio Tinto subsidiary ERA to restart uranium mining in Jabiru, is unsettling. Australia's track record on cultural sites is poor. He cannot rule out another Juukan Gorge.

Fabian realises that it is really urgent, now, to steer a course of conservation. Antarctica, the remains of Gondwanaland, the oldest place on the planet, the beating heart of the living biosphere, the fulcrum around which land masses have been revolving, mutating, drifting, colliding, folding and unfolding for so long, is rapidly melting. With unthinkable consequences. Nobody knows for sure, opinion is shifting towards earlier climate tipping point domino effects, such as the shutting off of the AMOC and the Gulf Stream by mid century. Causing a rapid onset of coastal inundation. He strengthens his resolve to go to Kakadu at the earliest opportunity, to talk to the people on the ground.

CHAPTER 36

Over the past few decades, North America has experienced a gradual rise in average temperatures. This warming trend is consistent with global climate change and has led to more frequent and intense heat waves, especially during the summer months. Many regions in North America have seen shifts in precipitation patterns. Some areas have experienced increased rainfall and more frequent heavy downpours, leading to flooding. Conversely, other regions have faced prolonged droughts, impacting water resources and agriculture.

The Atlantic hurricane season has seen an increase in the frequency and intensity of hurricanes. These storms bring devastating impacts, including strong winds, storm surges, and heavy rainfall, causing significant damage to coastal communities. Several areas in North America, especially in the western United States and Canada, have seen an increase in the frequency and severity of wildfires. Hotter and drier conditions have created a conducive environment for wildfires to spread rapidly, leading to significant ecological and economic consequences. The Arctic region, which impacts the climate of North America, has experienced a reduction in sea ice extent and thickness. This melting ice contributes to rising sea levels and can influence weather patterns, including the jet stream, which affects weather events in North America.

The increasing absorption of carbon dioxide by the

world's oceans has resulted in ocean acidification. This can have adverse effects on marine ecosystems and fisheries in North American coastal regions. California's coastal waters are acidifying twice as fast as the rest of the oceans, as recent studies show. And some of California's most important seafood, including lobster, squid and crab, are becoming increasingly vulnerable.

Joanna has been fascinated by the ocean and its diverse species of flora and fauna, as long as she can remember. Growing up in California, she spent countless hours exploring tide pools, snorkelling in kelp forests. Her love for the ocean was influenced by her family's affinity for the water. Growing up near the Californian coast, she naturally gravitates to the beach, and surfing quickly becomes her favourite pastime. Every chance she gets, Joanna heads to the beach with her board, delighting in learning to catch waves. Feeling the addiction that most surfers experience, she immerses herself in the beauty and power of the ocean.

Through surfing, Joanna develops a profound connection with the ocean. The feeling of gliding on the waves is like dancing on water. Being in sync with the rhythm of the ocean inspires a sense of awe and wonder. She becomes acutely aware of the delicate balance of marine ecosystems and their vulnerability to human impacts. She loves the feeling of being close to the dolphins, knowing that they are the master surfers. Her passion for the ocean inspires her to pursue a career in marine biology.

After leaving school, she continues her surfing adventures, and yet she cannot help noticing the challenges facing the plants and animals in the seas off the coast. This is her domain, at one with the ocean and the wind, alive to the beauty of blue sky reflected in sparkling waters, in love with the dolphins whom she occasionally sees in the different surf spots she frequents. Like many of her generation, she deplores the effects of plastic pollution, the decline of marine species, and the impact of warming waters on coastal environments. She knows that climate change is anthropogenic, caused by humans.

These experiences spark a curiosity to gain a better understanding of the complex interactions within ocean ecosystems. Being motivated to take action to protect the environment, Joanna realises that a career in marine biology will allow her to contribute to ocean conservation. She is determined to play a role in the restoration of habitats and species in the face of an alarming set of changes in the coastal waters noticed by the whole surfing community.

Joanna's deep surfing connection is part of her sense of responsibility to the ocean. She wants to contribute to the scientific knowledge that underpins effective conservation strategies. She focuses on the interconnectedness of marine life, gaining an understanding that simple systems form the complexity of the beautiful, magical environment of the sea.

Inspired by her local surf community, she becomes a passionate advocate for ocean conservation to friends and family, and anyone who will listen over a morning coffee at her local cafe hangout. She is determined to protect the oceans she loves, ensuring that future generations can enjoy the beauty and wonders of the sea that first inspired her in childhood.

She attends the State University of California, enrolled in marine biology. As an undergraduate, she immerses herself in the coursework as well as being out in the field. She participates in research projects, gaining hands-on experience, both in the field at the beaches, and out in the boats as an observer, as well as time spent in laboratory analysis. She definitely knows what she prefers, and that is to spend as much time as possible out on the ocean.

Joanna gets involved in a project studying the impact of climate change on California's kelp forests. She spends a couple of summers along the rugged coastline, collecting data on kelp populations, monitoring water temperatures, and studying the interactions between marine species in this critical ecosystem. Her research also encompasses understanding the effects of human activities on endangered species, the sea otters and blue whales, off the coast.

She actively participates in conferences, workshops, and seminars, where she engages in networking and communication within an international community. These networking opportunities lead to collaborations with other marine biology students, allowing her to gain exposure to a wide range of research techniques and perspectives. At one online conference, she encounters members of an Australian team conducting research on the Great Barrier Reef. With so many interests in common, they exchange contact details. Even at the initial online meeting, she feels an immediate sense of simpatico, particularly as the Australians have all learnt to dive, providing tantalising accounts of their experiences underwater in the clear waters off the Queensland coast.

She discovers that there are surfers among them, and within a few weeks, with a real feeling of camaraderie, she is invited to participate as a volunteer in a project in the Australian winter, coinciding with the summer vacation in the US, studying the effects of coral bleaching and the decline of sea grasses first hand.

The Coral Sea Marine Park is one of Australia's most spectacular treasures. The park covers a huge area off the coast of Queensland beyond the Great Barrier Reef, and is one of the world's largest. It has a diverse array of habitats, such as coral reefs, sandy cays, deep sea plains and canyons.

After bus hopping to Queensland, via towns with exotic names such as Waramungu and Camooweal, Jonas makes it to Townsville, a little weary from the long journey across a seemingly endless landscape of diverse country, with a variety of vegetation, sparse grasses and trees growing over ancient sedimentary basin, past mineral rocky outcrops, giving way to the tropical vegetation as the bus rolls into Townsville, situated on the coastal strip.

Jonas checks into the Coral Reef Backpackers in the city, a short walk from the Greyhound terminal. After a reasonably comfortable night in an eight bed dorm, the following morning he sets out to explore, walking along the Strand, the shore looking out over the inner shoals to Magnetic Island. He is pleased and amused to learn that the swimming conditions at the beach are pleasant, with no marine stingers in the waters and no crocodiles on the beaches today.

He rests up, in preparation for turning up the following day at the Centre of Excellence for Coral Reef Studies. It's on the university campus at the foot of Mount Stuart, and he finds his way by bus, at the appointed time, to a project induction meeting. He enters the building and follows the signed directions to the meeting room, just a little apprehensive.

He takes a seat with half a dozen other young people, patiently waiting for the woman at the front of the room, who is clearly an organiser. 'We'll just wait for a few more minutes before we start.'

Jonas observes the young woman who walks in the door asking the room at large 'Marine conservation project induction?' in an American accent.

The organiser takes the floor. 'Hello everyone, my name is Rhonda Jones, and this is to welcome you and prepare you to join the Inshore Coral, Seagrass and Water Quality Program. I think we have all six people here, so let's get started. We'll have a couple of days visiting sites around Townsville, then we'll be sending you out to various locations. We've arranged transport to the dock, and there'll be a boat, so we'll start with a safety briefing.

The weather in Townsville, Queensland in July is generally mild and sunny. The average temperature is 23 degrees Centigrade or 73 degrees Fahrenheit for our friend from California, and the humidity is also relatively low, which is good because water moisture on your skin makes you feel much hotter, as we are located in the tropics. Off the coast is the Coral Sea, the prevailing winds are southeasterlies, so you can look forward to mild weather and sunny days. The beaches are also a great place to relax and enjoy the sunshine.

Weather conditions for the day, light and variable winds with smooth seas, and sea temperatures average around twenty two, so good swimming weather. You'll need protective clothing and sunscreen. You'll be able to snorkel or dive, and we'll cover the specific safety procedures including what to do if you get stung. Then there is a run down on the species that you may encounter on the reef, including what to do if you see a shark. A really important aspect of this work is of course environmental protection, and as you are all engaged in marine research, there's no need to remind you how important this is. We'll provide useful local instructions for protecting the reef, disposal of waste and how to avoid touching the coral.

There'll be a team leader with you for the first few days to ensure that you have a safe and enjoyable experience monitoring inshore reefs. The sea level around Australia's coastline was approximately 100 to 120 metres lower once. As a result, Australia's continental shelf was exposed during the last glacial period. In the present-day, many areas where coral grows used to be exposed land.

A couple of tips for staying safe. Stay hydrated and wear sunscreen. Always be aware of your surroundings. Following the guidelines to ensure that you have a safe and enjoyable experience monitoring coral health, water quality and seagrass on our inshore reefs. Now I suggest you get to know each other because you'll be working together for the next four weeks, so morning tea is available in the foyer. Pick up your name badges on your way out, and introduce yourselves.'

There's an excited buzz of conversation, as the volunteers avail themselves of coffee, tea, fruit, biscuits and cake. Joanna has a lot of self confidence, she introduces herself, and after a few exchanges comes to rest next to Jonas, who is naturally shy. ' Hi. Joanna, from California, this is an amazing opportunity to get involved and learn from the Australian experience. They have some of the best research. What brings you here?'

'Hi Joanna, I am Jonas. My interest is marine organisms and ecosystems. Why are you here?' Jonas is pleased to talk to the healthy, tanned young woman in front of him.

'Oh, same. To learn more about marine habitats and species. I've been monitoring water quality and seagrasses off the Californian coast. Tell me more.' Joanna is natural, warm and encouraging, which immediately brings Jonas out of his shell.

'Me too. I've been monitoring water quality and seagrasses off the Norwegian coast. And looking at how we can maintain species diversity.' Jonas looks at his companion with interest.

'Lots in common, then.' says Joanna, deciding that her first impression of the man with the untidy blonde hair is a good one.

CHAPTER 37

Anna jumps up from her desk at the art deco John Gorton Building, close to Old Parliament House. She literally runs to Sandra's office, and bursts in without waiting for an invitation. 'Sandra, you've got to see this.'

Sandra looks up, with a discouraging look. It is not done for a junior staff member to burst into the office of a senior public servant. However she waits to see what kind of information would cause Anna to act outside of service behavioural norms.

'I've just looked at the Kakadu directory on google photos.' If Anna is hoping for immediate comprehension she is disappointed.

Sandra responds. 'Yes, photos and videos from official visits by Parks Canberra officers to Kakadu National Park.

Anna struggles to put what she has found into a form that would be easy to understand for a pre cloud computing generation. 'Alex took a video when he was at Gunlom. Photos and videos are automatically uploaded to the cloud when there is an available network. This means that you don't have to remember to manually upload it from your phone. Which is useful if you are in a location where the mobile signal is poor, or non-existent out in the field. Or there is something wrong with the phone's network connection.'

'Cut to the chase, Anna.' Sandra is looking blank, as she tries to make sense of what she's just heard.

'I just looked at a second video that Alex took out

at Gunlom Falls. The first is a close up of the stairs, and the barriers, and closure notices in the car park. I nearly deleted the second, because the front image is blurry rocks. But I didn't. You've got to see it. I've got it on my desktop monitor.' Anna rushes back to her desk and presses play, with Sandra following her. They stand back and watch a bad video with shaky visuals unfold.

There is a view of rocks, then the camera is moved to focus on a young man, clearly angry. 'You can't back out. Just because some abo says he doesn't want to do business with us. And if you tell anyone about this deal, you'll regret it.' A second voice is heard. 'Don't threaten me, Greg. I've got the phone camera rolling. I don't want any part of Top End Tourism.' The video becomes very shaky, however it is clear that the first man is coming closer to the camera. His voice is heard. 'Give me your phone.' The sound of a scuffle is recorded while the visual looks like it is a hand, then a pocket. Then a splash is heard, then the video ends.

Sandra and Anna look at each other, both clearly stunned. Sandra finds her voice. 'So do we know when this video was taken, Anna?' Anna's reply is barely audible. 'Parks mobile phone video cameras always have an automatic time stamp on all photos and videos. It was taken at 18:02 pm on Wednesday the 19th of July. And it automatically uploaded to Parks google drive at 03:40 am on Thursday the 20th of July. '

The two women are silent for a minute. Then Sandra speaks. 'Can you download this video, take a copy then upload it to the encrypted executive drive. Then delete it permanently from the Kakadu drive. Please do not disclose this to anyone else. I'll be contacting the

Northern Territory Police, once I've shown it to the management team. And go home for the rest of the day.'

Anna is only too happy to comply. She downloads the video to her local PC, then uploads it to the encrypted executive team drive, naming it Gunlom Falls, and emails Sandra that she's done as asked. She gathers up her laptop backpack, and heads to the cafeteria for a latte.

When she gets Anna's email, Sandra heads straight to the office of the Director of National Parks. They agree that the best thing is to contact NT Police immediately.

'Isabelle Felham?'

'Sandra Murphy, Parks Australia. Isabelle, I have a video that was automatically uploaded from Alex Petersen's mobile phone, which you'll want to see. I can give you encrypted access to the video with a link to Parks secure server, and a temporary VPN user id and password over the phone. But you'll have to promise to access the video from behind the NT Police network security perimeter. Then I presume you have an encrypted location for videos that are used in evidence. Then let me know when you have it, and we'll delete it from our secure storage.'

'Sandra, why the cloak and dagger? What's on the video?' Isabelle is incredulous.

'We believe it is a video taken as Alex fell into the pool at Gunlom Falls. His mobile phone automatically uploaded the video to the google drive Parks uses for Kakadu fieldwork purposes. And one of his colleagues just had occasion to access it this morning. It is definitely from Alex's phone, and it is Alex's voice, plus another male voice. It was taken about an hour after he met Bob Wilson. And was uploaded in the early hours of the morning. Presumably when the phone had network access. I am going to send you a link now, to download a VPN client. And by separate email, another link to access the video once you've got the Virtual Private Network up and running. And I'm going to give you the temporary user id and password now. Do you have a pen and paper? 'Sandra's instructions are crisp and clear.

'Go ahead, Sandra. I do have pen and paper.' Isabelle is glad that she always has a yellow post-it notes sticker pad at the ready.

A few minutes later, she heads for Rachel's desk and invites her to her office. Isabelle explains what she has just learned, and asks for Rachel's help. 'I am supposed to download a client VPN from this link. What's that when it's at home?' She turns her laptop around so that Rachel has access, indicating the yellow sticker stuck to the keyboard. That's the log in. And when you've got it up and running, I've got a link to the video in an email from Sandra.

Rachel busies herself at the tasks, while Sandra just sits silently, watching. It's only about ten minutes later that Rachel copies the video link from Sandra's email. She downloads it, then opens it with a default mp4 player. She turns the laptop at an angle so they can both clearly see the video which is playing in full screen mode. At the end of the two minute clip, Rachel asks. 'Do you want to see it again?'

'Yes.' Isabelle's eyes are glued to the screen. At the end, she speaks slowly. 'So Alex had his work mobile with him. And we did not get a forensics report about it, because it was not on his body. How did we miss that? Bob Wilson told us that he had a mobile phone when he was in Perron's car. We only got call records from his personal mobile phone, which he left in his hire car in Darwin.'

Rachel responds. 'Don't beat yourself up about it. There can only be one reason why the Scene of Crime team did not find it up at the pools. Crowther took it with him. So they were standing on rocks above the pool. Crowther was threatening Alex, because he refused to be part of Top End Tourism. Clearly whatever Bob said to him brought him to his senses. He was scared by Crowther's angry response, so he took out his mobile phone to record that he was being threatened. Crowther grabbed the phone from him because he did not want to be recorded. He grabbed the phone, and either Alex accidentally fell, or Crowther pushed him into the water. And left him there to drown.'

The two women say nothing, until Isabelle breaks the silence, with more than a note of anger in her voice. 'Pull Crowther in, and we interview Perron again.'

'I'll get the paperwork moving,' Rachel responds.

'I'll go to the Super now, to tell him what's really been going on.' Isabelle is already heading for the door.

CHAPTER 38

After another round of interviews, it does not take Isabelle long to complete her written report.

'Greg Crowther is an individual who was known to the deceased, Alex Petersen. They attended Darwin High School at the same time. In the past couple of years, Crowther has been associated with a Darwin resident, Barry Perron, who has been under surveillance by DOCS for the past few months, after a tip off from Indonesian Police that he is connected with a boat sailing from Bali to Darwin, that may have been carrying methamphetamine and MDMA. DOCS have discovered that Perron is known to Interpol from suspected illegal activities in Northern Italy and Switzerland.

Crowther was in contact with the deceased in connection with Perron's new business venture, Top End Tourism. From ASIC company records, Perron is a co-director of this company. The other director is a Malaysian individual, Lim Rajendran. Anecdotal evidence indicates that the co-director is associated with Malaysian company Persekitaran. Enquiries reveal that Persekitaran has been involved in illegal logging activities in many countries, and has recently branched into tourism. At interview, Crowther said that he had been promised a 'piece of the action' by Perron. Crowther said he had asked the deceased, Alex Petersen, if he could persuade Mirarr traditional owners to support Top End Tourism's proposal for the redevelopment of Jabiru.

Crowther was shown a video that uploaded to a Parks Australia's cloud storage automatically. There is very little mobile phone coverage in Kakadu National Park, certainly none at ground level. The video reveals that the mobile phone was removed from the deceased's possession at the pools above Gunlom Falls. Subsequent investigation reveals that the phone automatically connected to the Telstra mobile network when the vehicle driven by Crowther was within range of Darwin mobile phone towers. It is believed that Crowther had the deceased's mobile phone in the vehicle he was driving, a Holden Calais four wheel drive loaned to him by Perron for its backroads capability. It is likely that Crowther disposed of the phone in Darwin. The phone was turned off at 3.50 am, just before Crowther says he arrived back at his home. This is consistent with Darwin red light cameras that captured Perron's vehicle between 3.40 and 3.45.

The video shows a part of a heated argument between Crowther and the deceased at Gunlom Falls. It is clear that Crowther told Perron that the deceased had decision making capacity, and the power of recommendation to the Kakadu Board, in his role with Parks Australia. Anecdotal evidence from a Canberra colleague and friend of the deceased, indicates that Crowther told the deceased he was involved with a Persekitaran deal with the NT government for a hotel deal in Jabiru.

A second interview with Perron suggests that he, Perron, was introduced to Persekitaran through a mutual contact, looking for a Darwin based partner. He set up a company, Top End Tourism, with a second director introduced to him by Persekitaran. He was under the impression that Persekitaran had already successfully lobbied the NT government to establish tourist facilities at Jabiru, and that they required a local presence.

Crowther alleged that Perron told him that there was a Persekitaran deal, and that Top End Tourism would receive money to pay Traditional Owners for new facilities at Gunlom Falls in return for their support. Crowther offered to connect Perron with the deceased, in his role with Parks Australia, as the intermediary with the Jawoyn people.

Crowther also alleged that Perron informed Persekitaran that Kakadu traditional owners would jump at the offer of new infrastructure and cash payments. He said that negotiations had been in the pipeline before the deceased made his trip to Darwin on Parks business, to attend a Kakadu Board meeting. Crowther said that he thought that the deceased had the kind of influence required to negotiate with Traditional Owners.

During the meeting with the deceased at Fiddler's Green, Crowther informed Alex Petersen that Perron had told him that Persekitaran and Top End Tourism had obtained the support of the NT Government for a deal with the Jabiru Corporation, owned by local indigenous people.

Crowther said he informed the deceased that Perron had told him there were funds available to offer new infrastructure and money to traditional owners of Kakadu. He also said that Perron claimed Persekitaran already had an agreement with ERA, as well as the NT government. Crowther said he passed this information on to the deceased, who had agreed to being the intermediary with the Jawoyn people while they were seated at the bar of the Fiddler.

When interviewed, Perron denied all of these assertions. He, Perron, revealed that he was offered a joint venture in Top End Tourism by Persekitaran to provide local administration, and that after this came up in casual conversation over a beer, Crowther asked him if he could be involved. He denied that he had told Crowther that Persekitaran had a deal with either the NT Government or Rio Tinto subsidiary ERA. It may be that Perron offered Crowther a financial incentive to gain the support of Kakadu traditional owners through the deceased, although both Perron and Crowther denied this.

Forensic analysis indicates that fingerprints from both Crowther and the deceased have been found in Perron's vehicle. In consideration of the delay between the struggle with the deceased and when Crowther drove back to Darwin, it is not clear why Crowther waited so long. However an empty whisky bottle was found under the passenger side front seat of Perron's vehicle. The only prints found on the bottle were Crowther's. It may be that Crowther drank a quantity of whisky, then slept in the vehicle prior to driving back to Darwin, which would explain the delay.

Crowther states that the deceased suggested to start negotiating with traditional owners by offering money for upgraded parking and facilities at Gunlom Falls, and moving the offending stairs to a new location, and that he was convinced his offer would be successful.

Crowther and the deceased drove to Gunlom Falls, arriving around 3.45 pm. Crowther states that he left the keys with the deceased while he walked up to Gunlom Falls, to avoid being present when Alex Petersen met with Bob Wilson. When asked why Bob Wilson said that the deceased arrived after 4 pm, Crowther stated that Petersen had wanted to look at the adjacent campground while waiting for Wilson to arrive.

Bob Wilson, a Jawoyn traditional owner, stated that after showing the deceased the new stairs, he refused to discuss an overture made to him by the deceased, to support a deal in return for infrastructure repairs. Alex Petersen appeared to have realised from his conversation with Wilson that negotiations would be very difficult, and would certainly jeopardise his career, if discovered by his employer Parks Australia. The video evidence makes it clear that the deceased told Crowther he had changed his mind, and did not want to be involved. This angered Crowther, and the video clip clearly shows he is threatening the deceased. When the deceased tells Crowther he is capturing the conversation on video, Crowther moves towards the deceased.

While the video is very disjointed and does not clearly show what happened, it is evident that there is a struggle over possession of the phone. A splash is briefly heard just before the camera is switched off. It is inferred that this is when the deceased entered the water. According to Crowther, the deceased was fine when he left him. He now admits that he took the deceased's mobile phone. Crowther states that the deceased got out of the pool and was standing on the edge when he left him to descend to the carpark.

The view of this report is that the deceased lost his balance and fell into the water caused by Crowther grabbing his mobile phone. It is possible that he hit his head as he fell. It is also possible that Crowther subsequently hit him on the head with a rock, which he then disposed of in another location. No blood traces were found anywhere at Gunlom Falls. It is most likely that he did not get out of the water. He subsequently drowned. While the exact circumstances of the deceased's head wound and drowning are still not known, it is clear that Crowther's actions were a primary cause of Alex Petersen's death.

The recommendation is for charges to be laid, either of murder or manslaughter, depending on further investigation. There are also grounds to consider laying charges of criminal negligence, failure to provide assistance, reckless endangerment, and theft of a mobile phone.'

CHAPTER 39

Fabian arrives in Darwin at 2 am, on the only direct flight from Denpasar Airport in Bali to Darwin International. He passes by new comfortable lounges, airport shops and food outlets. People have rated Darwin as very new looking, clean and friendly, with a strong police presence on casual patrol. He passes by the sign 'The Defence Department advises the public that these essential works will ensure the integrity of runways and taxiways and the safety of aircraft and passengers. '

He is glad to find Melanie waiting for him in the arrivals hall. 'So sorry to make you come and collect me at this hour.'

'No problem, Fabian, just glad you've arrived safely.' Melanie is indeed relieved that Fabian has kept his promise. 'We can definitely do with some good advice. The car's this way.'

'So what's with the airport upgrades?'

'The airport shares runways with the Royal Australian Air Force's RAAF Darwin base. They are resurfacing runways. The investment is because it's a military asset, as well as for the NT public airport.' Melanie responds.

Next morning, over cereal, bacon and eggs, fruit and coffee, Melanie explains that she's organised a road trip to Kakadu, starting with a meeting at Jabiru, a kind of unofficial council. 'You've heard of Graham Saunders, the journalist? Well he's around, researching what the

Federal and NT Governments have cooked up for Middle Arm. And I've invited Bob Wilson, a Jawoyn traditional owner, and my cousin Ben Mooney, a Jabiru community constable. We've decided to fill you in on the murder of a Parks Australia employee at Gunlom Falls. '

Fabian looks alarmed. 'A murder? That's unbelievable? What happened?'

'It's a long story, Fabian. And we have a long drive to Jabiru, so I'll tell you on the way. We're picking up Graham, he's coming with us, and we're meeting at Bob's office at the Community Centre early this afternoon. Ben will join us there. He is going to fill us in on the results of the police investigation. He says the official report from the Scene of Crime Examination team has been finalised, and the Coroner's report is due out. We're hoping that you'll be able to give us an idea of where we can go from here, given the complicated politics involved.

We'll also be able to give you an inside view of the state of play with the decommissioning of Ranger, and how traditional owners can steer the rehabilitation of the Kakadu environment.'

'Sounds good, Melanie. Let's get going. Give me ten minutes to get ready?'

'No problem.' Fabian and Melanie stack the breakfast dishes in the dishwasher, and head for their respective rooms to gather bags and laptops. Having collected Graham, once out on the road on the Arnhem Highway, the conversation turns to Middle Arm.

Fabian addresses Graham. 'So what do you think of the development of new port facilities for Darwin. I can see that the airport is being upgraded?'

'It's not hard to see that there is money and influence driving government initiatives in the Northern Territory. I'd say there are two main sources, mining industry money and the influence of the US government on Defence, and the AUKUS deal.' Graham decides he won't pull his punches with the likeable Swiss. 'There are new moves afoot to expand the US military presence here.' Graham rolls out his explanations easily.

He's just finished writing this view as an article for an online news media. 'The governments have a hidden agenda of facilitating gas expansion at Beetaloo Basin, to drive a new industrial hub in Darwin. They are being persuaded by mining companies and other vested interests in the Northern Territory that the only way to grow the economy is by rapid industrialisation, and that means investment in old carbon intensive industries, such as petrochemicals, plastics, massive agricultural production, and new gas liquefaction plants to take advantage of global energy shortages. The NT government is influenced by wealthy companies, and the Federal government has persuaded itself that short term dirty industries can fund a green transition, some time in the future.'

'But that is awful.' Fabian's honest Swiss Republican heart is offended by such a dishonest policy and twisted economic rationalism. 'Surely the governments know that this won't work. That there is a huge time penalty for not addressing climate change, and the further the Titanic sails towards the iceberg, the harder it is to turn the ship away.'

'No. The NT government is known to accept backhanders. And the feds were climate deniers until a few years ago. Now they have been out of power so long, they are either naive or deliberately blind to how they are being manipulated by corporate greenwashing. In any case, they have a poor environmental track record, because they have been historically focussed on getting a share of the action out of the wealthy companies.' Graham decides it's too late to tell anything but the unvarnished truth.

Melanie turns the conversation to Kakadu. After all that's why Fabian is here. 'Of course, you've been following the debacle of Parks Australia building stairs right beside a Jawoyn cultural site. '

'Yes, of course,' replies Fabian. 'The NT Supreme Court found the Commonwealth immune from liability because of the wording of the legislation, so Parks Australia could not be held liable. Any updates on that situation, by the way?'

She provides a brief outline of progress to date. 'The High Court of Australia has agreed to hear an appeal by Traditional Owners, the Aboriginal Areas Protection Authority and the Northern Land Council . A long slow process, in other words.' Melanie proceeds to tell Fabian what has happened at Gunlom Falls, and how she discovered the body of Alex Petersen, someone she'd distantly known from school days, at the instigation of her cousin, Ben. She relates how suspicion had fallen on Bob Wilson, as the last person to see him alive. She follows this by an account of the astronomy course, and the field trip with Mara, Grace and Jonas coming forward to the NT Police about what they'd heard and seen. And how she and Graham had seen Alex with another man who'd also been to Darwin High School, at the bar of the Fiddler.

Her story draws to a close. 'That's as much as I know. I think that we're going to get an update from Ben. Hopefully, the coroner's report goes public today, so he can tell us what he knows. And I think Bob Wilson has some information for us too, extra information that he did not tell the police.'

'Bound to be an interesting meeting,' Graham remarks.

'This is like something out of a book.' Fabian sounds incredulous. 'Happening in real life. Tell me about the country we are passing through. It looks like eucalypt open forests and woodlands over grassland. Is this what you call savannah? '

Melanie responds. Yes, this is one of the many micro climates and ecosystems in the Territory. If you look carefully you'll probably spot some kangaroos grazing. Or maybe some wild donkeys and cattle.

'Part of a wildlife management challenge, I presume.' Fabian is thinking about his brief to review the World Heritage status. They drive through the country in comfortable silence, all three busy with private thoughts. They pass over the South Alligator River, observing the subtle variations in vegetation, on the lookout for kangaroos on the side of the road.

Melanie breaks the silence by turning on NITV radio. An interview with a Greens senator is being broadcast. 'Senator, what do you think about the US-Chinese relations? You seem to be well informed on international affairs?

'Well, I think it's not too late to defuse conflict. China has significant economic problems that are cutting its growth. They have an ageing population and sky high youth unemployment. Like the west, they also have high debt levels in both the private and public sectors. State capitalism is too restrictive, and private investment is falling. In the US, there are more and more banks exposed to rises in interest rates. The global economy is shifting away from the US dollar as countries explore alternative currencies and financial systems.Hopefully a shift towards a more sustainable future.'

'How do you see this situation affecting Australia's future?'

'The situation in Australia is that we've just jumped into bed with the US and the UK, committing to the AUKUS deal and high spending on nuclear energy for decades to come. The Labor Party is firmly committed to the AUKUS nuclear-powered submarines, contrary to public opinion, which is that they are not worth the $368 billion and rising price tag.' The senator pauses for a moment.

The interviewer continues. 'So why does Australia want to be tied to the UK and the US? The UK is in trouble, and the US dollar is shaky, which means that Australia may be spending a lot of money in countries where the industrial base is increasingly unstable?'

The senator tries to clarify the political position. 'I think a lot of people are worried that Australia is being drawn into someone else's conflict. Currently it's a cold war, and the Greens are hoping that it does not spill over into a hot war. That's why it is a worrying sign that the Australian government is not telling the Australian public about the growing commitment to US military assets here in the Top End. The US Air Force is building a mission planning and operation centre in Darwin. You'd have to be naive not to ask what else they are not telling us.

The proposed second port of Darwin would be a costly, carbon intensive industrial development, spending billions on fracking for gas and petrochemicals, despite the climate emergency. Behind this plan are wealthy vested interests. They are not spending their own money. The federal government has committed $1.5 billion in taxpayer money. It's a cynical scheme to reap profits based on a short term energy crisis while the world is transitioning to renewables. Burning fossil fuels is killing people by droughts, fires, floods and mega storms.'

The three people listening in the vehicle remain silent, finding nothing further to say about the seriousness of the situation.

Half an hour later, Melanie pulls into the carpark at the Community Centre. Bob Wilson comes out to meet them, and ushers them to his office. 'I'll give Ben a call to let him know you're here.'

'Tell him to give us half an hour to discuss Kakadu business,' Melanie adds.

After introductions, they settle in with a cup of tea. Fabian gets straight to the point. 'So what can be done to clean up the Kakadu environment. Invasive species are increasingly threatening the native ecosystems.'

'It's simple. Traditional owners need to drive the rehabilitation programs. Our young people have to get the park ranger jobs. Currently they're going elsewhere. There needs to be training programs. Parks Australia has to start listening to us. After all, we've always maintained this country, and we understand it. We know how it works, and what it needs to rebalance itself.

Right now Parks have too much power, and they think they have all the answers. As well as not employing our young people, they are not working with the natural environment in a sustainable way. When it comes to tourism, it's as though they are driven by politics, not by understanding how to care for this country. This is not going to change in a hurry. They may mean well,however they are run by the bureaucracy in Canberra.'

'So what do you see as the answer, Bob?' Fabian appreciates his candour, even while he wants a solution to the environmental degradation, so he can recommend that Kakadu Park keeps its World Heritage status.

'Traditional owners have to get our own money, so we can pay for what has to be done. That's the way we can work with Parks. Right now, they are not letting the funds committed by the federal government be spent. Until they decide what they want to do with them. What is supposed to be an equal partnership between us and them is anything but. We have to get some kind of eco tourism deal going. I was hoping it would be in a deal with the Jabiru Corporation. The problem is that anything that mob wants to do has to be approved by the federal and NT governments, and the mining company. You know ERA is a subsidiary of Rio Tinto.'

'Go on Bob, tell Fabian what you think can be done.' Melanie steers the conversation onto a solution.

'Well, if we work with Jabiru Corporation on an ecotourism deal, with a proper renewable energy hub, not a couple of solar panels and a diesel generator. If we can make money out of energy and eco tourism, we can upgrade the infrastructure for all the beautiful places that the tourists love to go, the waterfalls, the rock art sites. We can save them with enough funds to run proper environment restoration projects that are effective. We know how to manage this country. We know how to get rid of the invasive weeds and animals. We figure that a sustainability hub in Jabiru could be a technology demonstration centre of excellence of how to manage the environment, generate clean energy, provide a starting point for new clean technologies, and make it a tourist attraction as well.

Mirarr people don't just want luxury hotels and glamping and rich tourists who can pretend they are in the wilderness, we want a model eco destination where everyone can come. And learn about how to care for country, how to generate and store renewable energy, without using big battery banks made of so called green metals, mined by exploitation of people in Africa and South America, with terrible working conditions. We need partners to work with who share our vision, not companies that just want quick profits. We know how to work together, and we want to work with people who want to work with us.' Bob is suddenly silent, worrying he has said too much.

Fabian is impressed. 'I know you can do it, Bob. That is a great set of aspirations. And you are right, there is plenty of technology that does not require exploitation of people and corrupted national governments. What I am hearing is that you need partners who have solved the clean technology and ecotourism problems. Sounds like you just need start-up funding to start rolling, to put rehabilitation programs in place run by local people. And you have to be able to collect funds to maintain the infrastructure, while you are cleaning up the park. And you have to run the programs, because the government has clearly demonstrated that they cannot.'

'Yes, that is what we want.' Bob concurs. 'We just have to find a way to do it.'

'I think you can do it. And I think you have to start with ecotourism and renewable energy partners. And some funding of your own. Once you get started, the government money can be spent in a useful way, such as maintaining roads and tourist campsites. So you have to make it a condition of establishing the partnerships, that they provide up front finance for the rehabilitation works, with no expectation of any control of how the money is spent. I can advise you. It's not beyond my remit with UNESCO, and I am happy to help any way I can, including helping you to find the partners, and setting up the terms of investment.'

'OK. That's good. While you were talking I was thinking about something that Alex Petersen told me at the end of our conversation out at Gunlom. I did not tell NT Police, but he was honest with me about the deal he was hoping to get involved with. He told me that he knew Persekitaran were behind Top End Tourism. And that he'd heard from a friend in Canberra that they were intending to outsource the whole tourist development deal. I told him he was dreaming. Because it would have to be approved by Parks Australia, the NT government, the Jabiru Corporation and ERA. So he said he'd heard that Persekitaran had approached a company called Yang Sehat Resorts, based in Bali, who were really interested. And that Persekitaran was indeed in discussion with the NT Government.'

Fabian and Bob look at each other, struck by the same thought at the same time. Fabian is the first to speak. 'So why not approach Yang Sehat Resorts directly, find out what they have to offer. Cut out the middle men.'

'Good idea. If we can get through the red tape and we can get a request for proposals together, will you give us some feedback? It will have to be watertight to get the buy-in of the governments.' The wheels are clearly turning in Bob's mind.

'Of course, only too happy to help.' Fabian's response is immediate and sincere.

Thanks Fabian, appreciated.' Bob sees Ben pull up outside, and goes to greet him.

'Hello Ben.' Melanie greets her cousin. 'How are you doing?'

'Hi Melanie. Sorry about sending you to find the body. Are you over it?' Ben nonetheless has a cheeky grin.

'Oh I think so, it's nearly two weeks now. So what have you got to tell us, what's the inside story. Have they found out what happened?'

'They have.' Ben sounds triumphant as he turns his attention to the two men also in the room.

'Fabian, Graham, this is my cousin Ben, a community constable based here in Jabiru.'

'Nice to meet you, Ben.' Graham offers his hand.

'Yes, good to meet you Ben. Fabian Bertrand from UNESCO. ' Fabian also introduces himself.

'So Ben. What do you know about Alex Petersen's death.' Melanie gets straight to the point.

'OK. The formal press release says that he died after an altercation with Darwin man, Greg Crowther.'

'Aha! Graham, that's the bloke he was talking to at Fiddlers.' Melanie turns to Graham. 'So what happened. How did they find out?'

Ben tells the story of the relationship between Crowther and Perron, and Perron's joint business venture with Persekitaran, Top End Tourism. He doesn't know about Perron's drug connections, but he has heard about the automatic upload of the video to google photos, as the phone connected to the mobile network on the outskirts of Darwin. 'The phone hasn't been found. They think that Crowther disposed of it before he got home, because it was turned off about 15 minutes after the upload. He probably just took the SIM out, smashed it, and tossed it in a municipal bin. He wouldn't take the chance of it being found with what he thought was the only copy of an incriminating video. '

'So Alex and Greg were discussing the potential for Top End Tourism getting a piece of the action with the Jabiru Corporation and the NT government. Surely Alex would have realised that any deal would have to pass Parks Australia as well as the mining company, ERA? Why would he want to get involved with Persekitaran? '

'I heard that Alex had got involved with a woman staffer at the Indonesian embassy, the niece of a Balinese eco tourism entrepreneur, who'd been approached by Persekitaran. He would have done some research, and found that Persekitaran is dodgy, while the aunt's company, Yang Sehat Resorts has an excellent reputation, and a practical track record. I think he might have decided to find out more about what was happening. Starting with talking to Bob Wilson. He probably had some kind of idea of dealing directly with Dewi Pratama, through her niece. And he probably went along with Crowther to find out more about whether Persekitaran had actually already done a deal with the NT government.'

'Even before he talked to Bob, he must have known he'd have problems dealing with Top End Tourism. Pity he did not back out before going to Gunlom Falls.'

'I suppose he was persuaded by Crowther, who would really want the meeting to go ahead, because he had been boasting about Alex's influence to Perron. Maybe Alex didn't want to let him down. Maybe he did not realise just how dodgy the deal was until he spent a few hours driving from Darwin with him.'

'Or maybe he was just looking for the best deal he could get, and did not really care, as long as he was in the loop.'

'Do you know whether they've arrested Crowther?' Graham asks.

'Yes, this morning. They've charged him on suspicion of murder, though he'll probably get off with manslaughter.' Ben says.

'So that's that.' says Graham. 'I feel sorry for Alex Petersen. Have they got anything on Perron?'

'No. He's wriggled out of everything. Claims he knew nothing about Crowther's activities, that he had just mentioned there might be a tour guide job, not a business partnership.'

'He got off scot free, then?' Melanie asks.

'Seems like it.' says Ben.

CHAPTER 40

One of the directors of the Jabiru Corporation is speaking directly to the studio ABC camera. 'The Mirarr people have a vision for the region as a world leading ecologically sustainable, economically and socially vibrant community where traditional culture, people and the natural environment flourish. Mirarr people continue to live on our country as our ancestors have done since time immemorial. Jabiru is built on Mirarr country and we have a responsibility to look after the town and the people who live here. We have a Mirarr Vision for the town. We ask everyone who comes to do business to follow the Jabiru Master Plan as the guiding document for all future development and activity.

The visuals fade over to Graham, in the studio of the ABC Darwin. Looking directly at the camera, he reads from the autocue.

'Traditional Owners the Mirarr people have announced a new development for the Jabiru Corporation. This is progress on the implementation of the Master Plan developed with the Commonwealth and Territory Government, and Energy Resources of Australia. A joint statement released today is the result of several years of negotiations of how the federal government funding is to be spent. This is a significant development for Kakadu National Park to keep its World Heritage status. '

A clip of Jabiru town fades to the surrounding Kakadu National Park, followed by footage from Gunlom Falls panning to a spectacular sunset over the eucalypts, grasslands and wetlands. Shots of the waterfall travelling through the rocky escarpment, show it plunging to a base pool fringed with trees.

'A formal agreement was reached today with the Mirarr to establish a tenure arrangement for development that aligns with the Mirarr vision for the town, and traditional cultural values. The first round of proposals is being sought from interested parties for site development consistent with these values.

The Mirarr have agreed criteria for assessment of all development applications. This represents a significant milestone in the development of Jabiru as an Eco Prototype Centre of Excellence, to demonstrate to Australia and the world how town planning, residential building, waste recycling, renewable energy and storage can work in the context of the environmental rehabilitation of the Kakadu National Park.'

It's only a matter of months before Jabiru Corporation receives its first proposal under the new agreement, an eco tourism proposal by Dewi Pratama of Yang Sehat Resorts.

'The unique selling proposition of this ecotourism project is that we take care of the environment. This includes our supply chain, site location, building materials, and building designs. We take particular care of our interactions with the local community. We also bring local recycling and waste-to-energy technology that can be used by the town. We harvest the water we use, recycling grey water for the hotel gardens.

Our target market is broad. We engage with all age groups, demographics and interests. This includes older visitors who want comfortable accommodation, as well as young backpackers who are looking for budget community campground facilities.

We look to Kakadu Traditional Owners to provide culturally sensitive tour services, including transport, catering and tour guides. We encourage people to respectfully enjoy the natural environment of this World Heritage area. We are looking to establish partnerships with the Mirarr people for visual and performance art, including paintings, dance and music at purpose built hotel facilities. We are seeking to establish relationships with traditional owners to ensure that hotel visitors can be escorted and guided to the rock art sites, and the natural environment of the beautiful Kakadu National Park.

We are committed to engaging with the local people in a skills development program, to ensure that there are real employment and business opportunities for local people, not just for the hotel. The aim is to facilitate a growth in sustainable business to support a vibrant community sustaining economic growth for all, to ensure the long-term viability of Jabiru.

To this end we intend to be actively involved with a community renewable energy microhub for the whole town, with solid state hydrogen storage. We are happy to co-invest with state and federal governments in this project, as part of the Jabiru Eco Centre of Excellence to demonstrate new clean, green technologies.

Our management team has deep experience in ecotourism on the island of Bali, Indonesia. As part of the evaluation process, we can offer inspection of our current operations at our flagship hotel in Bali, to see for yourself that we truly have the skills and experience necessary to successfully manage the development and the operations of the ecotourism project.

We can also offer a contribution to the visitor centre facilities. We are happy to support the development of culturally appropriate audio visual displays of the wonders of the Kakadu National Park.

In summary, our promise is to improve the town of Jabiru with recycling and renewable energy, while protecting the water supply and the natural aquifers. Our transport commitment is to use hydrogen vehicles for local transport, and for visiting the wonderful rock art, gorges, pools and waterfalls in the Park. All waste is recycled. Local people are to receive new skills and on-the-job training. The hotel conference and performance facilities are available to local people and local businesses, as well as hosting global ecotourism events. Marketing is to be developed in conjunction with Traditional Owners. A financial summary is attached, detailed figures available on request.

By highlighting these benefits, we are hoping you are convinced that our facilities can attract potential investors. We believe our ecotourism project is a worthwhile investment.'

Bob is in the office of the Jabiru Corporation, meeting with key stakeholders. 'So we've had a proposal for a tourism eco hotel from Bali. Have you read it? What do you think?'

'Yes, I have.' Bob pauses, considering how to frame his reply. 'Better get the ball rolling. Contact the lady. Take her up on her offer to visit her eco tourism development in Bali. Talk to the people working there. Go and see the recycling plant. See what kind of renewable energy they've got going, how they do energy storage. Then if you like what you see, ask for the financial details, and get the Commonwealth and the Northern Territory governments, and ERA involved. Otherwise you might get fobbed off with some dodgy deal the government wants to sell you. '

CHAPTER 41

Mara moves around her room, packing her belongings. Suitcase organised, with things she'll need near the top, she packs her laptop into her backpack, does one final look around. She finds her shampoo and conditioner still in the shower, and with a sigh, unzips the suitcase and deposits them in her bathroom bag, zips up the case again, and heads out the door to the bus stop. By this time, she is comfortable with Darwin's public transport system, so she takes the local bus to the airport. She has plenty of time, her flight leaves early afternoon for Brisbane. This was not her original plan, which was to take buses around the country. She realises this is not practical, because she only has another week until she has to fly home. 'Next time I come to Australia, it'll be for three months.'

Mara enjoys the feeling of rising above the everyday events, to fly above the clouds. She is fascinated by the wide vistas of the country below. Having a window seat, she views the coastal trees, grasslands and wetlands suddenly giving way to the waters of the Gulf of Carpentaria. Crossing the coastline again, scored by channels of water courses, she looks down at the characteristic oxbow patterns in the low-lying plains as the rivers of the channel country meander their way to the sea. They are just skirting the interior of the continent, as the aircraft flies along the shortest route in three dimensional space, known as the great circle. While map projections make the Earth look flat, the great circle path is the shortest path between two locations distant from each other, given the curvature of the surface. She loves being reminded that the planet is part of space. The flight covers a distance of over two thousand five hundred kilometres, with a flight time around three and a half hours. The country below unfolds at a slow pace. Mara doesn't mind. She is really enjoying seeing the topography of the landscape below.

The Gulf Country gives way to a flat dry country sparse of vegetation, the edge of Australia's dry central interior, before the aircraft heads southward along the ridges of the Great Dividing Range that separates central Australia from the coastal floodplain running the length of eastern Australia. This blocks the flow of moist air coming from the Coral Sea, creating rain over the range, and reducing the amount of rainfall in inland regions. As they cross to the coastal plain, Mara can see greenery below her. They pass forests, ranges and gullies. The terrain below suddenly stops being wild country, giving way to the subdivisions of pastoral properties, towns and river systems and more forests. Suddenly they are passing over smaller farms and then suburbs. On the last stage of the flight, they cross the coast to fly along the edge of the bay, taking a right hand turn towards Brisbane Airport, landing into the westerly wind, crossing the mouth of the wide river winding below.

Mara emerges from the aircraft, making her way to the station, and the train that takes her to the Gold Coast, her first stop travelling south to Sydney. Deciding she does not want to increase her carbon footprint, she is going by bus first to Byron Bay, and then on to Sydney. Tonight's accommodation is an airbnb at Currumbin

Beach. This turns out to be a beautiful surf beach with white sand, blue waves, and friendly local cafes. After a morning swim in the flagged area monitored by local surf lifesavers, she sits on the deck and watches the world go by at a leisurely pace. Fortunately, she's decided to stay another night, and makes the most of the happy, relaxed atmosphere at the beach, dining at the surf club, overlooking the sweeping view north to the islands of Moreton Bay, with the waves rolling into the beach a mere twenty metres from her table.

The next day is also fine. At the local newsagency she picks up the morning paper, before sampling an acai bowl and an excellent coffee from the cafe next door. She learns from one of the friendly locals that the tourist promotion for Queensland is 'beautiful one day, perfect the next'. She admits that as far as the weather goes, it fits. She catches the local bus to the Tweed Heads Bus Station, in good time to board an afternoon coach to Byron Bay. Views of the wide vistas of the hinterland plateau, the green caldera of the extinct volcano Wollumbin, is truly spectacular. They motor past a distant Mount Warning, named by the British explorer James Cook in 1770, a towering rocky shape over deep green bush and farming country, while glimpses of the ocean appear towards the east. It's a short journey, and despite having heard that it is past its best, she likes the look of the town, with its air of arts and crafts with a hint of hippy.

Finding her overnight accommodation not far from the centre of town, she walks to the bay, a long coastline that ranges from a headland escarpment, complete with lighthouse, along sandy beaches that disappear in the direction from which she's come. She enjoys the next few days, walking to the lighthouse and along beach tracks, originally made by the Arakwal people, the Bundjalung of Byron Bay. There are plenty of cafes and art galleries that appeal to her, and she does her best to avoid the more touristy spots.

She reads a little about the history of the Bundjalung nation. She learns that the clan group inhabit the coastal areas and hinterland north of the Gold Coast as far south as Yamba. According to oral traditions the first settlers arrived by sea, landing at Goanna Headland, just south of Evans Head, one of the most easterly points on mainland Australia. Goanna Headland is an amazing topical feature of the terrain where bushland meets the ocean. It is believed to be a mythological place of origin of the Bundjalung Nation inhabited for at least 12,000 years, since the end of the last ice age. Mara wonders if this was a migration where people had to move inland as the seas rose over the continental shelf when the ice melted. She makes a decision to stop on her journey to Sydney.

Starting out early in the morning, she catches a bus that lets her off in the main street of the little town that has its origins in farming and fishing. After a good breakfast at a local cafe overlooking the Evans River estuary, she makes arrangements to leave her gear for the day, and having got directions and a map, starts the hike to Goanna Headland. Crossing the mouth of the Evans River, climbing the steep hill to the lookout, she can see why Evans Head is a haven for ocean lovers. The view is unique. Looking north, she can see not only the ocean, but the green ancient volcanic hinterland all the way north to Springbrook. She finds the path to Chinaman's Beach in the bay formation south of the headland, named for the nineteenth century Chinese settlers panning for alluvial gold who had their encampment at the beach. The day is wearing on, and she has to cover the five kilometres back to town. Having taken the precaution to find out about local taxi services before she set out, she gets a ride back to the bus stop, arriving in time for the overnight bus to Sydney.

The next morning, leaving her luggage at Central Station, she makes her way to the City of Sydney, crossing Hyde Park and the Domain to the beautiful classical nineteenth century Art Gallery of New South Wales. Finding the cafe, she sees her friend from the astronomy course, Grace, already waiting for her at an outdoor table in a sunny spot overlooking Woolloomooloo Bay.

'Hi Mara, great to see you again.' Grace rises from the table and gives her new friend a hug.

'You too, Grace. So glad we can meet again before I have to leave tomorrow.'

'Already? Too short! You'll have to come back again.' Grace turns towards her two companions at the table. 'Malcolm and Eka, colleagues at the Australian Museum. This is Mara. You've heard all about her, now here she is.' There are greetings all round, and Mara feels a genuine warmth from this group of Aussies.

'Any friend of Grace,' she responds, 'is bound to be interesting and wonderful.'

'Thanks for the compliment!' Eka responds sincerely.

'So have you enjoyed your time here?'

Mara proceeds to recount her adventures since she left Darwin, particularly the journey through Bundjalung territory. 'What a country. And how amazing are the first Australians? I have learned so much on this trip, I am so glad I took that endless flight from London, even though I did not think so at the time.'

Her companions laugh. They are all veterans of the interminable flights halfway around the world to get to the places that their ancestors called home. They queue for lunch, and choose from a tempting array of simple healthy food and beverages, before settling back to the table and continuing the conversations.

'So Mara,' Grace looks expectantly at her new friend.' Tell us all about Scotland, stone circles and the Standstill Moon that's happening next year.

'If I let you in on the secret, will you promise to visit Aberdeenshire stone circles to view the event?'

There is acceptance all round.

'OK. The major lunar standstill happens approximately every eighteen and a half years. Our ancestors knew them to be times when exceptional weather and tides happened, particularly flooding. Aberdeenshire has the greatest concentration of recumbent stone circles. They first appeared in the late Neolithic around 2,500 BC. They have a massive stone laid horizontally on its side in the southwestern or southern arc of the circle, flanked by the two tallest stones. The recumbent stones are huge and heavy. They were carefully levered into place and wedged up so that their upper surface is as level as possible.

The path of the sun is called the ecliptic, though it is us who are moving round the sun, not the other way around. So the backdrop we see in the night sky revolves too, as we move around the sun once a year. The Moon's orbit is inclined about 5° from the ecliptic, so we see it swing in a regular cycle against the backdrop of the night sky. This swing has northerly and southerly extremities, called standstill points. It takes eighteen and a half years for the cycle to repeat.

When the Moon is at its most southerly point, at high latitudes north, it rises and only just skims the horizon, which looks spectacular. The time of the next lunar standstill event will be September 2024 and March 2025. The ancients considered the lunar standstill so important that stone circles were purpose built, particularly in Aberdeenshire, where the latitude means that the Moon appears to touch the Earth. The recumbent and flanking stones form a viewing frame to view the rising and setting, and the moon appears to dip towards the recumbent stone. Spectacular. The ancients made cup marks, hollows in the stones, clustered at points where the major standstill moon rises or sets. Ten of the best stone circles are within striking distance of Aberdeen. '

CHAPTER 42

A notice appears at the Darwin Yacht Club. 'Sailing from Darwin to Indonesia. Looking for crew. Leaving Darwin before the end of August. Destination Bali. No experience necessary. Must have a valid passport, and enough money to prove you can pay for an airfare home. Customs and immigration has to be cleared the day before we sail.You don't have to have a visa into Indonesia, though it is advisable to get one before you leave. You can get it on arrival. Sailing is a lot cheaper and a lot more fun than planes. Email bazper69@gmail.com.

A newspaper report appears in the NT News.

'Three people were all arrested in Sydney in connection with the bust and face court over alleged links to an international methamphetamine syndicate. The AFP alleges a man from Darwin was running the Australian operations organised from Thailand.

A search warrant was executed at a man's Darwin home, where the Australian Federal Police seized thousands of dollars in cash, electronic devices, and steroids. An eight metre fishing boat was also seized from the Top End residence, believed to be an 'instrument of crime'. The man involved was not at the residence.

Criminals can flee to another country to try to evade justice. A Red Notice alerts police worldwide about internationally wanted fugitives. This is a request to law enforcement worldwide to locate and provisionally arrest a person pending extradition, surrender, or similar legal action. It is based on an arrest warrant or court order issued by the judicial authorities in the requesting country. Member countries apply their own laws in deciding whether to arrest a person. A Red Alert has been issued by NT Police Drugs and Organised Crime in Darwin containing identifying information for Barry Perron, alleged to be involved in money laundering and an amphetamines syndicate. The Red Notice is published by Interpol as an international alert for a wanted person. It is not an arrest warrant.'

CHAPTER 43

They've been walking for five days. First they go to the rocks where Kuralini's grandfather said there was a map of the water. There they find the directions in a clear rock art carving. And the whole extended family group heads in that direction the next morning. They are heading roughly towards where the sun is disappearing at the end of the day. It feels like a good move. At night the first red fire star is shining brightly, as does her rival, the nearby star also with fire magic. They line up the direction of travel, keeping the seven sisters in the right perspective, spotting landmarks that will guide them, so that the next day, they know which way to go. Sure enough, Kuralini's direction finding, helped by Yindi and Waru, is good.

They find the water hole in the rock on the second night. At night they listen to the stories told around the fire. And just before they go to sleep, they ask Kuralini if she thinks the ancestor spirits are still guiding them. She is sure they are. She thinks that the time is right, because the spirits are back where they were a long time ago. When Yindi and Waru ask her why she thinks this, she just says that she sees it in the night sky, and hears it in the creeks, and feels it in the rocks and the trees. After four days, when Yindi and Waru are beginning to think they'll never find their new home, Kuralini tells them that the next day they will find an old place where their ancestors once lived. There'll be trees all around, good water to drink, plentiful bush tucker, and plenty of kangaroos. They can stay there. No one lives there now.

When Yindi and Waru ask her how she knows this, she points out that the animals, the birds and the country are telling her, and she sees no signs of people.

The next morning, they break camp. After eating and cleaning up, they look towards the direction that they decided last night would take them to where they are going. Late that day, they see the grasslands stretching out in front of them, dotted with eucalypts and low wattles. The journey over the light sandy soil is easy going. They are on a low plain well above sea level. They see a long line of red rocks. They head towards them, and when the last light of sunset has almost disappeared, they come to a place near a freshwater rivulet emerging out of the cliff, flowing to the sea. The water seeps along the edge of the red sand dunes, discharging onto the low tidal zone. From the low hill, they can see where they can collect water with their paperbark water carriers, and rock pools where they can spear fish. On the hilly plateau, they see open areas where they can camp. They've already seen several fat kangaroos. Yindi and Waru found a goanna for dinner along the way. Kuralini picked some gamba nuts. That evening, around the fire, they breathe a collective sigh of relief. They have found a new home, there is water and there is food. They have enough time to prepare before the wet season. Kuralini, at the fervent request of Yindi and Waru, says she will tell them one more story. 'It's a short one, because I am tired now.' She waits a few moments for dramatic effect, then begins.

'A long time ago, the sky beings decided to make a track across the universe to the land, the waters and the skies of our people. They did this so they could keep an eye on us, make sure we were respecting all the plants and animals, all the fish, the dugongs, the turtles and the dolphins. Sometimes they would send a message to show us what they wanted us to see. How to move around the country safely and where to find the water holes in the rocks. Sometimes they would send a message about the spirit beings of our ancestors, to let us know they are safe, and watching over us. You know those songs I taught you to sing. Well, I learned them from my grandfather. He taught me how to travel across the country. That's how I know which way to go. So now you know. Other songs tell us about creation beings and what they mean to us.

It's up there in the stars. Sometimes they show themselves to us, to remind us where we come from. Look up at the sky at night. Always remember the songs and what they tell us. We are the ones who have to work out what to do to keep ourselves and our country safe. Because one day, we'll have to remember the way again. And we can only find the way if everything is where it is supposed to be. Not only the night sky. The land, the sea, the animals, the grasses, the eucalypt trees, the bushes, the waterholes, and the streams that are always flowing. We have to care for them. And we have to watch as the sun and the moon say good morning and good night, and then we have to look at how the stars move at night across the sky. If we do that, we can always find our way home.'

ABOUT THE AUTHOR

Murray Stuart

I write about climate change because of the precariousness of our situation. Mystery Down Under is the first book in the series of mystery novels. The Universe of Blazingly Beautiful and Brilliant Souls connects us to our deep ancestral past.

THE UNIVERSE OF BLAZINGLY BEAUTIFUL AND BRILLIANT SOULS

Mystery Down Under is the first book of a climate fiction trilogy.

The second book in the series returns to Aberdeen, Scotland, and the Standstill Moon at the recumbent stone circles. The friends uncover the political deals to increase oil and gas mining in the North Sea. Back in Norway the whispers in the Sami community that China is funding a secret sea base near Murmansk, not far from Kirkenes on the Norwegian border, require investigation. Are Russians and Chinese interests planning to secretly deep sea mine for green minerals? The Sami people's network gets wind of a coastal nuclear submarine base, and the news soon gets back to Norway. Is there a link with spy operations conducted by low satellite technology?

Catch up with the Eco Centre of Excellence in Kakadu and the decommissioning of Ranger. Can't wait to find out? Sign up at www.goodbeachreads.com to be notified when Crosscurrents is released.

Crosscurrents

On the Black Isle, north of Inverness, the late 17th century bore witness to a land steeped in ancient traditions and simmering political tensions. Amidst the rolling hills and eerie stone circles, where whispers of the Neolithic past lingered in the air, the clash between the old ways and the encroaching modernity was palpable. Here, the struggle for Scottish independence against the English aristocracy, who leveraged their wealth from the slave trade to depopulate the Highlands, was a battle fought not just on the fields, but in the hearts of its people.

The young musician, son of a traditional farmer, embodied this resistance. His family, staunch supporters of the Jacobite cause, held fast to their roots and their land, defying the allure of quick riches offered by the absentee Sassenach landlords. After a droving trip to Inverness with a few of the family's hardy black cattle, Jamie was on his way home. A day that ended with the hopeful melodies of his fiddle started with a return journey, broken at Muir-of-Ord, where he sought companionship and relaxation with a few glasses of ale at the local hostelry. The warmth of the fire and the camaraderie of locals, acquaintances, and strangers offered a brief respite from the grim economic and social realities of the time.

As he stepped out into the night, the crisp air a sharp contrast to the dim, smoky warmth of the inn, he was unaware of the danger that lurked in the shadows. The path home led him past a deep, marshy pond, its still waters a dark mirror to the starless sky above. It was here, amidst the reeds and the whispers of the night, that someone struck a cruel blow. The sudden, sharp pain of a dagger's strike, a flash of agony, and the world spun as he was rolled into the marshy ground. Consciousness faded as the cold, reedy water embraced him, pulling him into the depths of the dark, dark pool, silencing the notes of Jamie Matheson Mackenzie's life forever.

The Black Isle is an ancient place. Recent archaeological digs have unearthed a rich tapestry of Bronze Age artifacts, revealing much about the ancient inhabitants of the area. Tarradale has always been a particularly significant community over it's 8,000 years of history. Among notable finds are evidence of extensive Bronze Age settlement, including agricultural tools and ceremonial items such as tinned bronze flat axes and socketed axes. These discoveries highlight the advanced metalworking skills and the complex societal structures of the past.

Excavations at sites like Balvattie Farm revealed intriguing features such as large ring ditches and pit complexes, which suggest the presence of significant communal activities. Aerial surveys have identified numerous crop marks and circular features, some of which were excavated to reveal a mix of ditches and pits, likely used for seasonal gatherings, astronomical observation and festivals. The Black Isle's archaeological record provides a vital link to Scotland's ancient past, offering insights into the lives and farming practices of early Highlanders.